Boundaries and Hulls of Euclidean Graphs

of Euclidean Graphs

From Theory to Practice

Boundaries and Hulls of Euclidean Graphs
From Theory to Practice

Ahcène Bounceur
Madani Bezoui
Reinhardt Euler

CRC Press
Taylor & Francis Group
Boca Raton London New York

CRC Press is an imprint of the
Taylor & Francis Group, an **informa** business

CRC Press
Taylor & Francis Group
6000 Broken Sound Parkway NW, Suite 300
Boca Raton, FL 33487-2742

Printed on acid-free paper
Version Date: 20180719

International Standard Book Number-13: 978-1-138-04891-1 (Hardback)

Library of Congress Cataloging-in-Publication Data

Names: Bounceur, Ahcène, author. | Bezoui, Madani, author. | Euler, Reinhardt, 1950- author.
Title: Boundaries and hulls of Euclidean graphs : from theory to practice / Ahcène Bounceur, Madani Bezoui, Reinhardt Euler.
Description: Boca Raton, Florida : CRC Press, [2018] | Includes bibliographical references.
Identifiers: LCCN 2018017134| ISBN 9781138048911 (hardback : alk. paper) | ISBN 9781315169897 (ebook)
Subjects: LCSH: Graph theory.
Classification: LCC QA166 .B6845 2018 | DDC 511/.5--dc23
LC record available at https://lccn.loc.gov/2018017134

Visit the Taylor & Francis Web site at
http://www.taylorandfrancis.com

and the CRC Press Web site at
http://www.crcpress.com

Dedicated to

El-hadi, Nadia, Kahina, Inès and Lina
(Ahcène)

M^d Ouslimane, Rezkia, Samira, Nilia and Louiza
(Madani)

Jeannine and Daniel
(Reinhardt)

Contents

Acknowledgments

We would like to express our gratitude to the *Agence Nationale de la Recherche* (ANR) for having funded the project PERSEPTEUR, which greatly motivated the work presented in this book.

We are grateful to Ali Benzerbadj, Kechar Bouabdellah, Audrey Cerqueus, Laurent Clavier, Pierre Combeau, Mohammad Hammoudeh, Loïc Lagadec, Farid Lalem, Abdelkader Laouid, Massinissa Lounis, Olivier Marc, Alain Plantec, Bernard Pottier, Marc Sevaux, Abdelkamel Tari and Rodolphe Vauzelle for their contribution; and to Julien Arguillat, Christian Fotsing, Salvador Mir and Basel Solaiman for their assistance and encouragement.

We are indebted to Callum Fraser, Sarfraz Khan and the editorial team for their valuable advice throughout the preparation of this book.

Our special thanks go to our families for their constant and invaluable support.

Preface

This book is intended for readers who need to determine the boundary of a set of nodes that can be represented as a network or generally as a connected Euclidean graph. It can be used for research as well as high-level university education.

New and recent algorithms are presented allowing to find boundary nodes and polygon hulls in connected Euclidean graphs. This kind of algorithm can be used in any centralized or distributed system arising in medical applications, data mining, Big Data, Wireless Sensor Networks (WSNs), Smart-cities, Smart-grids, Internet of Things (IoT), etc.

The main issue in finding boundary nodes of a connected Euclidean graph is the definition of the boundary itself. What are the boundary nodes in a connected Euclidean graph? We can give a first answer by means of a simple example, in which we try to modify Jarvis' convex hull algorithm by adding the constraint that in each iteration, only the nodes that are connected to the current node can be considered. Beyond the polygon representing the boundary itself, we will be interested in the associated polygon hull which is the area circumscribed by such a polygon. In this book, you will find three definitions and models of such a polygon hull that are related to the type of the considered graph.

Basic notions of graph theory, a description of special types of graphs and some relations with computational geometry are the subject of the first chapter. Particular emphasis is laid on polygons, for which some new notions and concepts are introduced which are related to their form and the way to visit their vertices and angles. This is essential for the characterization of *interior* and *exterior* polygons or the study of *pseudo-polygons* and useful for the understanding of the next chapters. You will then discover a list of algorithms existing in the literature and dealing with graph and shape reconstruction. They include the approaches based on Voronoï diagrams, Delaunay triangulations and Alpha Shapes.

You will then encounter the *Least-Polar-angle Connected Node (LPCN)*-algorithm, the true kernel of this book, which represents our proposed algorithm to find the boundary nodes of a connected Euclidean graph, where we determine in each iteration the connected node forming the minimum polar angle with the current and the previously found nodes. This process starts from a boundary node and, for the sake of coordination, we suggest to select the node with minimum x-coordinate. There exist many versions of *LPCN* that can be used according to the type of the considered graph. A particular version of *LPCN*, called *Reset and Restart LPCN (RRLPCN)* will be presented as an evolution of *LPCN* in which the starting node can be any node of the graph.

Since the presented algorithms can also be used for distributed or autonomous communicating systems like computers, cars, UAVs, people or smartphones, etc., you will find an introduction into distributed programming, followed by the distributed versions of all those algorithms presented in their centralized form in the previous chapter, especially the *Distributed-LPCN (D-LPCN)* and the *Distributed-RRLPCN (D-RRLPCN)* algorithms. Even though finding the node with a minimum value, especially the minimum x-coordinate, is easy within centralized systems, this is not the case for distributed systems. This problem is also called *Leader Election,* issue for which we present 5 new and recent algorithms: *Minimum Finding (MinFind), Local Optima to Global Optimum (LOGO), Branch Optima to Global Optimum (BrOGO), Dominating Tree Routing (DoTRo),* and *Wait Before Starting (WBS).* The *D-LPCN* algorithm requires to start with a leader election algorithm. These algorithms can be combined to guarantee additional reliability like fault tolerance. In the case of *D-RRLPCN*, where the starting node is fixed manually, it is possible to use the leader election algorithm *WBS* to start the process automatically without having to fix the starting node a priori.

The reader will then discover a platform called *CupCarbon* which is a simulator of WSNs dedicated to Smart-cities and IoT. This platform is written in Java and available online as an open source software and as an executable whose source code can be consulted. It offers an ergonomic and easy-to-use interface for visualization allowing to develop and validate algorithms, to check and understand visually what happens during simulation. This simulator also offers the possibility to program those distributed systems, which represent wireless sensor nodes deployed in a city. You can also develop centralized algorithms that you can directly integrate into the source code. It is this simulator that played a significant role in writing this book because it has been used to implement and validate the majority of the presented algorithms.

In Chapter 6, you will find a presentation of 5 applications in which the search for the boundary nodes is essential to solve the addressed problems. This chapter does not just show how to apply the presented algorithms, but also points to some specific situations that can arise and the theoretical concepts that can be used to solve them. The first application concerns the determination of the boundary nodes of a WSN using *LPCN* and *D-LPCN* algorithms as well as a comparison between them. The second application shows a specificity of these two algorithms that allows to detect faulty nodes during the determination of the boundary. The third application deals with gaps and voids especially in the case of WSNs. The fourth application shows how to use *LPCN* to extract the different clusters as polygons from a set of two-dimensional data. The *LPCN* algorithm allows the extraction of clusters that have a certain distance between them because we need to connect points without the risk of connecting clusters. The last application shows how to use *LPCN* to draw, in a binary image, the polygon hull of pixels of a zone-of-interest. This is possible because such a set of pixels can be represented as a connected Euclidean graph. Combining this method with the one presented in the fourth application allows to extract different parts of a fingerprint as a set of polygons that can be saved and used for comparison and the reconstruction of the initial image of the fingerprint.

Finally, you will find a discussion about angle graphs where we try to replace a Euclidean graph by a classical abstract graph, in which the weight of an edge is given by the values of the angles and known dynamically as a function of the visited nodes.

This book summarizes our main research as it has been conducted during the last 3 years at the interface of Graph and Network theory, Computational Geometry, Communication and Simulation. The results have been presented at high-level, international conferences and published in recognized journals of these fields.

Brest, March 9, 2018

Ahcène Bounceur
Ahcene.Bounceur@univ-brest.fr

Madani Bezoui
Mbezoui@univ-boumerdes.dz

Reinhardt Euler
Reinhardt.Euler@univ-brest.fr

Chapter 1

Fundamentals on graphs and computational geometry

Graphs are a powerful tool for decision-making and, beyond, an indispensable instrument for the representation and the modelisation of many real-world situations. In a graph, points called vertices may be pairwise connected by oriented or unoriented lines called arcs or edges, respectively, according to the problem treated. We can find an ever-increasing number of applications for both *'directed'* and *'undirected'* graphs: planning, organization, production management, networks, etc. It turns out that the concept of a *'finite, undirected graph'* will be sufficient for our purpose. For a comprehensive and more rigorous treatment of graph theory, we refer the interested reader to [7, 16].

In this first chapter, we introduce some basic definitions of graph theory together with those classes of graphs, which are of particular interest for our research work, and we briefly discuss their relation with computational geometry.

At the end of this chapter, the reader can find a definition of polygons and some new concepts introducing the notions of an interior, an exterior and a pseudo-polygon. The focus is on the angles and the way of visiting them at the polygon's vertices, allowing us to develop the idea of an angle-based method.

1.1 Basic definitions

Definition 1. A *graph* $G = (V, E)$ is given by a finite set of *vertices* (or *nodes*) $V = \{v_1, v_2, ..., v_n\}$ and a set of *edges* $E = \{e_1, e_2, ..., e_m\}$, i.e., a collection of subsets $\{u, v\}$ of V.

Example 1. Figure 1.1 exhibits a graph $G = (V, E)$ with:

$$
\begin{aligned}
V &= \{v_1, v_2, v_3, v_4, v_5, v_6\} \\
E &= \{e_1, e_2, e_3, e_4, e_5, e_6, e_7, e_8, e_9\} \\
&= \{\{v_1, v_2\}, \{v_2, v_3\}, \{v_3, v_4\}, \{v_4, v_6\}, \{v_1, v_5\}, \{v_5, v_6\}, \{v_1, v_6\}, \{v_3, v_5\}, \\
&\quad \{v_4, v_5\}\}.
\end{aligned}
$$

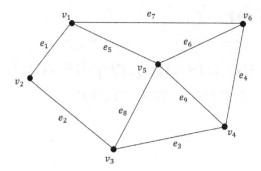

Figure 1.1: A graph $G = (V, E)$ with 6 vertices and 9 edges.

Definition 2. Two vertices u and v of a graph $G = (V, E)$ are called *adjacent*, if $e = \{u, v\}$ is an edge of G, such an edge is also said to be adjacent to its *end vertices* u and v.

Example 2. In Example 1, vertices v_1 and v_5 are adjacent, whereas v_2 and v_4, for instance, are non-adjacent. Also, the edge e_7 is adjacent to vertices v_1 and v_6.

Definition 3. The *order* of a graph is the number of its vertices.

Definition 4. The *degree* $d(v)$ of a vertex $v \in V$ is the number of edges of G containing v.

In Example 1 again, the order of G is 6, and $d(v_1) = 3, d(v_2) = 2, d(v_5) = 4$, for instance.

Property 1. The sum of the degrees of the vertices of a graph $G = (V, E)$ is equal to twice its number of edges.

$$\sum_{v \in V} d(v) = 2 \times |E| \qquad (1.1)$$

Definition 5. A *loop* of a graph G is an edge that connects a vertex to itself.

Definition 6. A graph is said to be *simple*, if it contains neither loops nor multiple edges, i.e., all edges are 2-element and distinct subsets of V. Unless otherwise stated, we suppose all graphs studied in this textbook to be simple.

1.2   Partial graphs and subgraphs

Definition 7. The graph $G' = (V, E')$ is a *partial* graph of $G = (V, E)$ if $E' \subseteq E$. We can say that such a G' is obtained from G by deleting some edges of E.

Definition 8. The graph $G' = (V', E')$ is a *subgraph* of $G = (V, E)$ if $V' \subseteq V$, and $E' \subseteq E$. We can also say that a subgraph of G is a graph resulting from the deletion of some vertices with all their adjacent edges and some additional edges.

Example 3. Figures 1.2(a) and (b) illustrate these definitions on the graph G presented as Example 1.

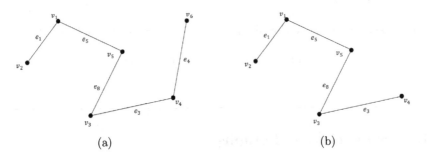

(a)   (b)

Figure 1.2: Examples of (a) a partial and (b) a subgraph of G.

1.3   Chains and cycles

Definition 9. A *chain* of *length* k in a graph G is a sequence $C_k = (v_1, v_2, ..., v_{k+1})$ of vertices of G such that v_i and v_{i+1} are adjacent for $i = 1, .., k$. We also say that the chain connects v_1 to v_{k+1}.

Definition 10. A chain is *elementary* or a *path* if its vertices are all distinct, *simple* if its edges are all distinct and *closed* if $v_1 = v_{k+1}$. A *cycle* is a simple and closed chain. We denote a path or cycle of length k by P_k or C_k, respectively.

Example 4. Figures 1.3(a) and (b) illustrate a chain and a cycle in the graph given as Example 1.

Definition 11. The *distance* between two vertices in a graph G is the length of the shortest path connecting them.

Definition 12. The *diameter* of G is the greatest distance between two of its vertices.

In Figure 1.3(a), the chain represented in bold is elementary and simple and thus a path.

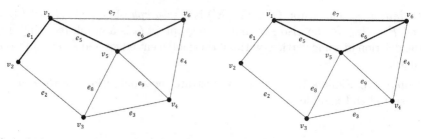

(a) A chain connecting vertices v_2 and v_6 (b) A cycle of length 3

Figure 1.3: A chain and a cycle.

1.4   Some classes of graphs

There are various types of graphs, depending on the disposition of the vertices and edges, the topology of the graph or the existence of specific properties. In the following, we briefly expose some basic classes of graphs.

Definition 13. A graph G is called *regular of degree k* (or *k-regular*), if all its vertices are of degree k, i.e., $\forall v_i \in V, \ d(v_i) = k$.

Definition 14. A graph G is *connected*, if there is a chain between any pair of its vertices. A maximal connected subgraph of G is a *connected component* of G.

Example 5. Figure 1.4 exhibits some examples of graphs having the mentioned properties.

Remark 1. The graph in Figure 1.4(a) is also a 3-regular graph.

Definition 15. A graph is *complete* if each of its vertices is adjacent to all the remaining ones, i.e., $\forall v_i, v_j \in V, \{v_i, v_j\} \in E$. K_n will denote the complete graph of order n.

Figure 1.5 illustrates the complete graph K_5.

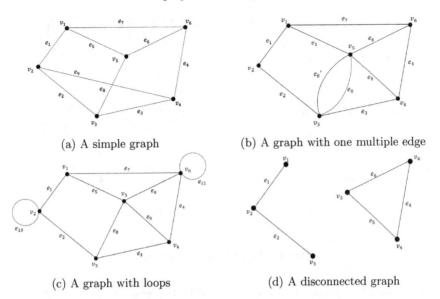

(a) A simple graph

(b) A graph with one multiple edge

(c) A graph with loops

(d) A disconnected graph

Figure 1.4: Some examples of graphs.

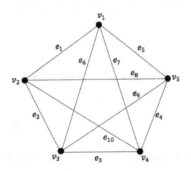

Figure 1.5: The complete graph K_5.

Definition 16. The *graph coloring* problem is the problem of assigning colors to vertices (or edges) of a graph under the constraint that two adjacent vertices (or edges) do not have the same color.

Definition 17. A graph $G = (V, E)$ is called *bipartite* if its vertex set V can be partitioned into V_1 and V_2 such that any edge of G has one endpoint in V_1 and the other in V_2, i.e., $\forall \{v_i, v_j\} \in E, (v_i \in V_1 \land v_j \in V_2) \oplus (v_i \in V_2 \land v_j \in V_1)$, where $\oplus$ represents the exclusive or (XOR). A bipartite graph with $|V_1| = m$ and $|V_2| = n$ containing all possible edges will be denoted by $K_{m,n}$.

Clearly, a bipartite graph can always be colored with 2 colors. Moreover, cycles of odd length allow for a useful characterization:

Theorem 1. [37] A graph is bipartite if and only if it does not contain an odd cycle.

Figure 1.6 shows two bipartite graphs.

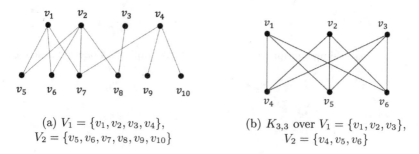

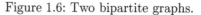

(a) $V_1 = \{v_1, v_2, v_3, v_4\}$,
$V_2 = \{v_5, v_6, v_7, v_8, v_9, v_{10}\}$

(b) $K_{3,3}$ over $V_1 = \{v_1, v_2, v_3\}$,
$V_2 = \{v_4, v_5, v_6\}$

Figure 1.6: Two bipartite graphs.

Definition 18. A *wheel of radius k*, denoted by W_k, consists of a cycle of length k and an additional vertex adjacent to all vertices of the cycle. A wheel of radius k will be denoted by W_k.

Example 6. Figure 1.7 exhibits wheels of radius 4 and 5.

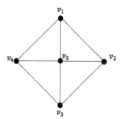

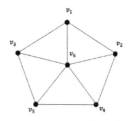

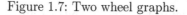

(a) W_4: A wheel of radius 4   (b) W_5: A wheel of radius 5

Figure 1.7: Two wheel graphs.

1.5   Hamiltonian graphs

Definition 19. A *Hamiltonian cycle (path)* of a graph G is a cycle (path) visiting every vertex of G exactly once.

A graph is called *Hamiltonian* if it contains a Hamiltonian cycle.

Theorem 2 ([16]). If in a simple graph of order $n \geq 3$ we have $d(v_i) + d(v_j) \geq n$ for every pair v_i, v_j of non-adjacent vertices, then G is Hamiltonian.

1.6 Planar graphs

Definition 20. A graph G is *planar* if it can be drawn on the plane in such a way that no two edges cross (i.e., intersect outside their end vertices). Such a representation is also called an *embedding* of G.

Example 7. See Figure 1.8 for an illustration of planar and non-planar graphs.

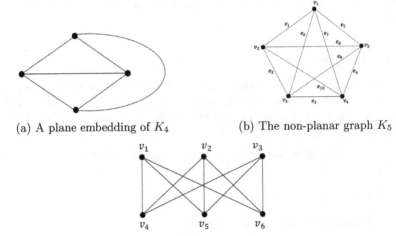

(a) A plane embedding of K_4 (b) The non-planar graph K_5

(c) The non-planar graph $K_{3,3}$

Figure 1.8: Planar and non-planar graphs.

Definition 21. A *face* of a planar graph is a region of the plane delimited by a set of edges and not containing other vertices or edges.

Theorem 3 (Euler, 1752). [16] Consider a planar graph G, whose number of vertices and edges is n and m, respectively. Then its number of faces f satisfies $f = m - n + 2$.

Definition 22. Two graphs G and G' are *homeomorphic*, if both can be obtained from the same graph by a subdivision of its edges.

Theorem 4 (Kuratowski, 1930). [75] A graph is non-planar if and only if it contains a subgraph homeomorphic to $K_{3,3}$ or K_5.

1.7   Trees

Definition 23. A *tree* is a connected graph without cycles.

Example 8. The graphs represented in Figures 1.2(a) and (b) are trees.

Definition 24. A not necessarily connected graph without cycles is called a *forest*.

Definition 25. In a tree, a vertex of degree at least 2 is an *internal* vertex in contrast to a vertex of degree 1, which is called an *external* or *outer vertex* or a *leaf*.

1.7.1   Properties

The concept of a tree is widely used in the fields of decision theory, data structures, graph drawing, etc. The importance of this type of graph has been the motivation for a thorough study of specific properties, some of which can be resumed as follows:

Theorem 5. The following statements are equivalent for any graph with n vertices:

1. G is a tree,

2. G is connected and has no cycle,

3. G has no cycle and has $n - 1$ edges,

4. G is connected and has $n - 1$ edges,

5. Each pair of distinct vertices of G is connected by a unique path.

1.7.2   Spanning trees

Definition 26. A *spanning tree* of a connected graph G is a partial graph of G representing a tree.

Spanning trees have been the subject of much research in various fields of computer science: algorithms and their complexity, network theory, etc.

1.7.3   Minimum spanning trees

The minimum spanning tree problem (MST) can be formulated as follows: given a real-valued weight-function w on the edges of a connected graph G, find a spanning tree of G of minimum total weight.

Example 9. Given the weighted graph presented in Figure 1.9, we would like to find a spanning tree of the graph which has minimum total weight[1].

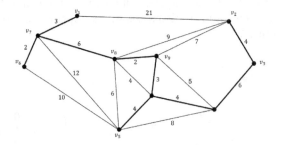

Figure 1.9: A minimum spanning tree.

The resulting minimum spanning tree is given by the following set of edges:

$$\{\{v_1, v_7\}, \{v_7, v_6\}, \{v_7, v_8\}, \{v_8, v_9\}, \{v_9, v_{10}\}, \{v_{10}, v_5\}, \{v_{10}, v_4\}, \{v_4, v_3\}, \{v_3, v_2\}\}$$

with total weight equal to 34.

Minimum spanning trees are used in image segmentation [4, 122], cluster analysis [91, 121, 125, 127, 128], time synchronisation problems in Wireless Sensor Networks [14], the efficient generation of minimum length multi-Wireless Sensor Networks [52], classification [69], routing problems in Wireless Sensor Networks [118], manifold learning [124], density estimation [81], data mining [68], pattern recognition [126], or machine learning [38]. To cope with the computational complexity arising within huge data sets, the authors in [126] present a minimum spanning tree algorithm on a complete graph of n points, based on a divide-and-conquer scheme to provide an approximate solution to MST.

The study of MST has started in 1926 with Borůvka who provided the first algorithmic solution of this problem by the Greedy algorithm [91], an achievement which later on developed into the theory of matroids which is nowadays considered as one of the cornerstones of combinatorial optimization.

1.8   Non-graphical representations of a graph

Information technology has revolutionized many fields, and it has contributed to solve problems that have remained unsolved for a long time, as is the case for several problems in graph theory. Solving such problems on a computer often requires an appropriate representation of the graph.

---

[1]Depending on the application, the weight may be a measure of the length of a route or the energy required to move between locations or the capacity of a line, etc.

1.8.1   Adjacency matrices

A simple graph $G = (V, E)$ can be represented by its *adjacency matrix A*, a symmetric $n \times n$-matrix, whose elements a_{ij} take the value 1 if the vertices v_i and v_j are adjacent, and 0 otherwise. An example of an adjacency matrix is given in Figure 1.10.

Figure 1.10: A graph and its adjacency matrix.

1.8.2   Adjacency lists

There is another way to represent a graph, called *adjacency list*, which enumerates for each vertex the adjacent vertices. An example of an adjacency list is given in Figure 1.11.

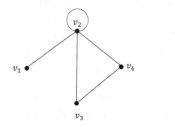 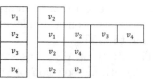

Figure 1.11: A graph and its row-wise adjacency list.

1.9   Computational geometry

At the beginning of this section, it is useful to recall the notion of a *geometric graph* which is a graph whose vertices or edges are associated with geometric objects or forms. This type of graph is widely studied in *geometric graph theory*, a special field of graph theory closely related to *Computational Geometry*, one of the oldest fields of computing with a history stretching back to antiquity. Since the problems we are going to tackle are at the heart of geometric graph theory, we feel appropriate to present some basic definitions.

1.9.1 Triangulations

The triangulation of planar point sets has been and still is actively studied for its wide range of applications, for instance in network theory, computer graphics and data mining.

The triangulation of a planar graph gives rise to several concepts of network analysis like: transitivity [58], clustering coefficients [62] and trigonal connectivity [8]. Becchetti et al. [9] utilize the number of triangles to identify spam pages in web designs and to measure content quality in social networks.

Definition 27. A *triangulation* T of a planar set of points S is a subdivision of the plane determined by a maximum set of non-crossing edges whose vertex set is S.

The notion of maximality here means that we cannot add an edge without crossing an existing one. Some properties specific to a triangulation of a finite set of points $S \subset \mathbb{R}^2$, given as a collection ζ of triangles, are

- $\bigcup_{T \in \zeta} T = conv(S)$, the convex hull of the points in S.

- $\bigcup_{T \in \zeta} V(T) = S$, i.e., the set of vertices of T is identical to S.

- For every distinct pair $T, U \in \zeta$, the intersection $T \cap U$ is either a common vertex, or a common edge, or empty.

Figure 1.12: Different triangulations of a planar point set.

There exist various criteria under which to construct a triangulation of a graph or a point set (see Figure 1.12). Of particular interest is that of Delaunay, which will be detailed in the following section.

1.9.2 Delaunay triangulations

This kind of triangulation is used to build the topological structure of data sets and named after *Boris Delaunay* for his work [40] published in 1934 and entitled "Sur la sphère vide. A la mémoire de Georges Voronoï", which can be translated as: "On the hollow sphere. In memory of Georges Voronoï". In this major dissertation, Delaunay presented a new triangulation procedure, a basic idea of which can be formulated as follows:

Lemma 1 (Delaunay's lemma [40]). *Let T a collection of tetrahedra which uniformly share the n-dimensional space being contiguous by $n-1$-dimensional integer faces and such that any limited domain has common points only with a finite number of these tetrahedra.*

Then the necessary and sufficient condition for no sphere circumscribed to such a tetrahedron to contain in its interior any vertex of any of these tetrahedra is that this takes place for each pair of two of these tetrahedra contiguous by a face with $n-1$ dimensions. That is to say, in each pair, the vertex of one of these tetrahedra is not inside the sphere circumscribed to the other and vice versa.

Definition 28. Let S be a set of points. A *Delaunay triangle* is a triangle with three points of S as its vertices such that there is no point of S in the strictly interior circumscribed circle.

Proposition 1 (Delaunay condition [39]). Three points $p, q, r \in S$ form a Delaunay triangle if and only if the circumcircle of these points contains no other point of S.

This concept is used in several fields, especially computer graphics and analytical geometry, more specifically surface reconstruction (Gopi [53]) and ad-hoc Wireless Sensor Networks (Li et al. [82]). A more popular method for Delaunay triangulations can be found in [78], where Lee et al. present two algorithms for the triangulation of a planar set of points: a divide-and-conquer procedure and an iterative algorithm.

We would like to mention at this place that a triangulation of a planar graph embedding can be used to obtain the polygon describing its outer face. It is one of the original features of the Least Polar-angle Connected Node (LPCN) algorithm to obtain this polygon in a direct way, i.e., without triangulation as it is done, for instance, in [57, 108].

Observation 1. A *Voronoï diagram* associated with a planar set of points $S = \{s_1, ..., s_n\}$ is a partition of the plane into cells $C_1, ..., C_n$, such that for $i = 1, ..., n$, C_i contains s_i and every point in C_i is closer to s_i than to any other point $s_j, j \neq i$. If we connect two points of S by an edge whenever the associated cells have a boundary in common, we obtain the corresponding Delaunay triangulation, as shown by Figure 1.13. Both concepts can even be shown to be *dual* to each other and they are fundamental for various applications. For further details we refer the interested reader to [50].

1.9.3   Planar straight-line graphs

As defined above, no two edges of a planar graph embedding are crossing outside their end vertices, even if the edges are drawn in curvilinear form.

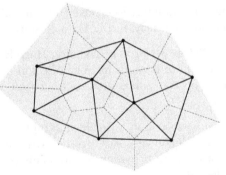

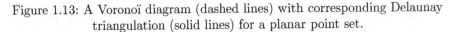

Figure 1.13: A Voronoï diagram (dashed lines) with corresponding Delaunay triangulation (solid lines) for a planar point set.

Within the context of geometric graphs, we may exclude such curvilinear forms as follows:

Definition 29. A *planar straight line graph (PSLG)* is a planar graph embedded in the plane using only straight-line edges.

Wagner (1936) [115] and, independently, Fáry (1948) [46] and Stein (1951) [106] have shown the following:

Theorem 6 (Fáry 1948 [46]). Every simple planar graph can be embedded in the plane using only straight-line edges.

Hence, for planar graphs there is no real loss in using only straight-line embeddings.

1.9.4 Euclidean graphs

Euclidean graphs have been used to model geometric problems in fields such as motion planning [35], circuit layout [103], or Wireless Sensor Networks [18, 28, 100].

Definition 30. The *Euclidean length* of a vector $p = (p_x, p_y)$ in the plane is given by $|p| = \sqrt{p_x^2 + p_y^2}$. We will denote by $dist(p,q)$ the Euclidean distance between two points p and q, i.e., the Euclidean length of their distance vector.

Definition 31. [123] A *Euclidean graph* is a straight-line embedding of a graph in the plane, the *length* of an edge being given by the Euclidean distance between its two endpoints, and its *weight* implicitly by its length.

Several notions of classical graph theory have to be adapted to the particular case of Euclidean graphs. We can cite:

- A minimum spanning tree of the complete Euclidean graph is called a *Euclidean minimum spanning tree*, and the MST naturally becomes the EMST. This problem is often encountered in road or transport networks, (when solving approximately Euclidean traveling salesman problems, for instance) and more recently, in Big Data clustering [86].

- The *Hadwiger-Nelson problem*, named after Hugo Hadwiger and Edward Nelson, asks for the minimum number of colors required to color the plane such that no two points at distance 1 from each other have the same color [41].

- The *shortest path problem* in a Euclidean graph. Given a set of polyhedral obstacles in a Euclidean space together with two points, the problem is to find a shortest path between the two points that does not cross any of the obstacles [61].

The notion of a *planar* Euclidean graph has to be redefined since Euclidean graphs are already embedded in the plane. It will be sufficient for us to distinguish between Euclidean graphs not containing any edge-crossing that we simply call *plane*, and general Euclidean graphs, i.e., those in which such crossings may occur.

1.10   Polygons and pseudo-polygons

1.10.1   Polygons

Definition 32. A *polygon P* is a sequence of line segments forming a closed loop. The line segments are the *edges*, and their intersections the *vertices* of the polygon.

Definition 33. A *simple polygon P* is a polygon having segments that do not intersect (see Figure 1.14). In terms of Euclidean graph theory, a simple polygon is a plane, elementary cycle.

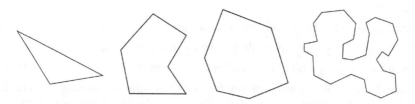

Figure 1.14: Examples of simple polygons.

Some of the problems to be introduced in the next chapter deal with the area occupied by a polygon in the plane. For this purpose, we just recall the method to follow for its calculation.

Let P be the simple polygon given by the vertices $v_1, v_2, v_3, \ldots, v_n$ with (v_i^x, v_i^y) representing the coordinates of the vertex v_i, for $i = 1, \ldots, n$.

Then the area of P can be calculated as follows:

$$Area_P = \frac{1}{2}[(v_1^x \times v_2^y) + (v_2^x \times v_3^y) + ... + (v_m^x \times v_1^y)]$$
$$- \frac{1}{2}[(v_1^y \times v_2^x) + (v_2^y \times v_3^x) + ... + (v_m^y \times v_1^x)] \qquad (1.2)$$

Definition 34. A *crossed polygon* P is a polygon having at least two segments that intersect (see Figure 1.15). In terms of Euclidean graph theory, a crossed polygon is a non-plane, elementary cycle, which contains at least one edge-crossing.

Figure 1.15: Examples of polygons.

Definition 35. An α-*loop* is a simple chain in which only the first and the last segment intersect outside their vertices.

Figure 1.16(a) shows three different α-loops.

Figure 1.16: Examples of α-loops.

Definition 36. An α-*polygon* is a crossed polygon where each edge-crossing results in an α-loop.

Figure 1.17(a) shows two α-polygons each containing a single α-loop, (b) shows an α-polygon containing two α-loops, and (c) illustrates another α-polygon containing three α-loops.

Note that α-polygons are not directly related to the applications considered in this textbook. A detailed study will be the subject of future research.

Definition 37. A *torn polygon* is a polygon having at least two vertices with the same coordinates.

Figure 1.18 shows an example of a torn polygon where the vertices A and A' have the same coordinates.

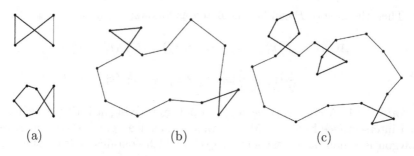

Figure 1.17: Examples of α-polygons.

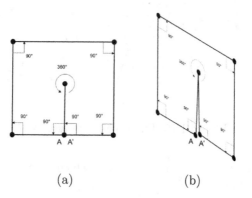

Figure 1.18: An example of a torn polygon (a) front view, (b) perspective view.

1.10.2 Pseudo-polygons

In the following, we will introduce a family of geometric objects, called *pseudo-polygons*, whose shape is close to that of a polygon.

Definition 38. Two polygons are *adjacent* if the interiors of their areas do not intersect and if they share at least one edge.

Definition 39. A *pseudo-polygon P* is a set of simple polygons in which each couple of polygons is connected by at most one chain in such a way that none of the polygons is subdivided into two adjacent polygons.

Figure 1.19 shows examples of shapes that are not pseudo-polygons.

Based on this definition, we can distinguish several types of pseudo-polygons.

Definition 40. An *octopus polygon* is a pseudo-polygon composed of one polygon and a number of chains that are connected to its vertices.

Definition 41. A *pseudo-polygon of type 1* is an *octopus polygon* the chains of which can intersect between them, but not with the edges of the polygon.

Figure 1.20(a) shows an example of a pseudo-polygon of type 1.

Figure 1.19: Examples of non-pseudo-polygons.

Definition 42. A *pseudo-polygon of type 2* is an *octopus polygon* the chains of which can intersect between them, as well as with the edges of the polygon.

Figure 1.20(b) shows an example of a pseudo-polygon of type 2.

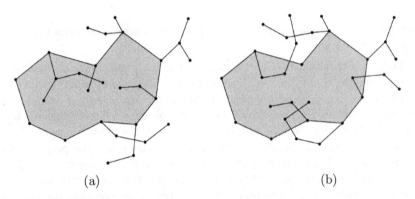

(a)   (b)

Figure 1.20: An octopus polygon of (a) type 1 and (b) type 2.

Definition 43. A *pseudo-polygon of type 3* is composed of several pseudo-polygons of type 1 and a number of chains connected to their vertices. The chains can intersect between them, but not with the edges of the polygons, which in turn cannot intersect between them.

Figure 1.21(a) shows an example of a pseudo-polygon of type 3.

Definition 44. A *pseudo-polygon of type 4* is composed of several pseudo-polygons of type 2 and a number of chains connected to their vertices. The chains can intersect between them, as well as with the edges of the polygons, which in turn cannot intersect between them.

Figure 1.21(b) shows an example of a pseudo-polygon of type 4.

Definition 45. The pseudo-polygons of type 3 and type 4 are called *generalized octopus polygons*.

Definition 46. A *tackled polygon* is a pseudo-polygon composed of a polygon and a chain connecting two vertices of the polygon without creating adjacent polygons.

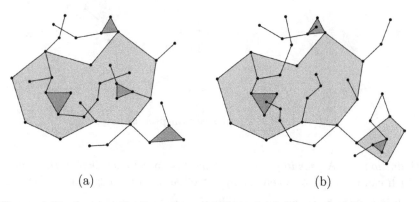

(a)   (b)

Figure 1.21: A generalized octopus polygon of (a) type 3 and (b) type 4.

Proposition 2. The chain of a tackled polygon intersects at least one of the edges of the polygon.

There are three possibilities to form a tackled polygon which depend on the way how the chain is added. The first one is obtained by starting the chain with an edge e_s inside the polygon and finishing with an edge e_f outside the polygon or inversely, as shown by Figure 1.22. We may call it a *pseudo-polygon of type 5*. The second one is obtained by starting the chain with an edge e_s inside the polygon and finishing with an edge e_f also inside the polygon. This case is shown by Figure 1.23(a) and we may speak of a *pseudo-polygon of type 6*. The third one, finally, is obtained by starting the chain with an edge e_s outside the polygon and finishing with an edge e_f also outside the polygon, as shown by Figure 1.24 and we may speak of a *pseudo-polygon of type 7*. In these figures, the non-adjacent polygons obtained after adding the chain are highlighted in gray.

Note that it is possible to have more than one intersection between the chain and the initial polygon. For example, Figure 1.23(b) shows another case for a tackled polygon of type 6.

1.11   Angles and visits

Angles and their visits are the notions to which we will pay particular attention in this section. This will be useful for the definition of different types of polygons as well as their characterization. Note that, in the following, we will not consider the case of α-polygons.

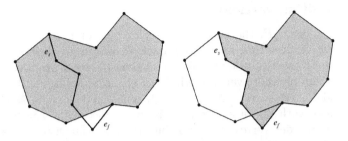

Figure 1.22: A tackled polygon (type 5).

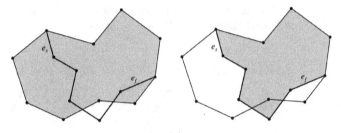

(a)

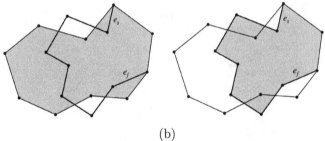

(b)

Figure 1.23: A tackled polygon (type 6).

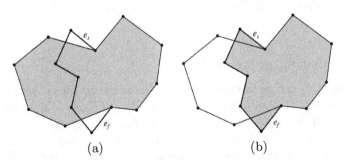

(a)   (b)

Figure 1.24: A tackled polygon (type 7).

1.11.1   Visiting vertices

Given a polygon P, the notion of a *vertex visit* of P can be defined as a sequence of vertices that are visited along P from a given *starting vertex* and where any vertex can be visited several times. From a given vertex, it is possible to visit only those vertices, which are connected to it. For illustration, let us consider three vertices A, B and C, connected as a chain as shown in Figure 1.25(a), with the objective to visit two vertices in a way to form an angle. If we consider vertex A as starting, then we can visit vertex B and then C, as shown in Figure 1.25(b), but we may also visit vertex B and then come back to A. If we consider vertex C as starting, then we can visit B and then A, as shown in Figure 1.25(c), or visit B and then C again. Finally, if we consider vertex B as starting, then we can visit either A and then B (see Figure 1.25(d)), or C and then B (see Figure 1.25(e)).

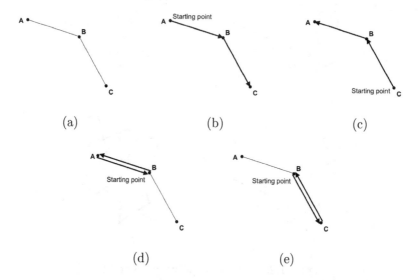

Figure 1.25: Visiting vertices in a chain of length 2.

1.11.2   Visiting polar angles

Visiting two vertices from a starting vertex in a chain, as shown previously, will lead to the visit of angles formed by them. Figure 1.26(a) shows two angles $\alpha = \widehat{ABC}$ and $\beta = \widehat{ABC}$ formed after visiting the vertices of the chain of Figure 1.25(a) from A to B to C. As we can see, these angles are illustrated in the same way but they are not the same. To avoid confusion and for consistency, throughout this book, we will work only with *polar* angles, i.e., angles that are formed in anti-clockwise order. Hence, on the basis of this rule, visiting the chain from A to B to C will lead to the angle α, as shown in Figure 1.26(b), and visiting the chain from C to B to A will lead to the angle β, as shown in Figure 1.26(c). In addition, a visited angle can be determined by the sequence of vertices that form it.

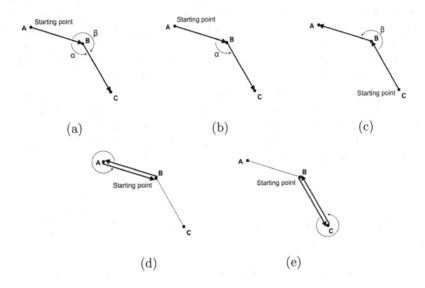

Figure 1.26: Angles formed by visiting three vertices of a chain.

Finally, if we visit from B either vertex A and then B again, or vertex C and then B again, we form an angle of 360°, as shown in Figure 1.26(d) and Figure 1.26(e).

1.11.3 Interior and exterior angles and polygons

In the following we will use the notion of *fictitious vertex* associated with a given vertex, which is a vertex having the same y-coordinate as the given one but a smaller x-coordinate. For the following examples, we will assume that the visit of the vertices starts from an angle formed by a fictitious vertex, its associated one, and one of its neighbors.

The introduction of the order of a vertex visit will help to determine which angle is formed by such a sequence. The visit of the vertices stops as soon as the first angle is visited a second time. Note that, before reaching the first visited angle, any point can be visited several times but each angle is visited only once.

Based on this rule, we will now define the type of an angle which can be either *interior* or *exterior*. These notions have a sense only when we work with polygons. In the case of chains, all angles are considered as exterior except if the chain is inside an interior polygon. In the next section we will define the notion of interior and exterior polygons, which will help to decide if an angle is interior or exterior. In case that an angle belongs to an interior polygon, it will be considered as an interior angle, otherwise, it will be considered as an exterior angle.

As an example, let us consider the polygon of Figure 1.27(a), where we try to visit the vertices based on their polar order and starting from vertex A. We obtain the polygon shown by Figure 1.27(b), where the vertices are visited following their exterior angles; we may thus speak of an *exterior* polygon. However, if we start from vertex B, we will obtain the same polygon, shown by Figure 1.27(c), but the vertices are now visited following their interior angles, and the polygon becomes an *interior* one. The main issue in determining interior and exterior polygons is how to characterize the interior and the exterior angles of a polygon. Figures 1.28(a) and (b) show that even if we start from the same angle as in Figures 1.27(a) and (b), we obtain two different types of polygons.

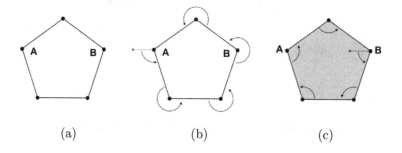

(a)   (b)   (c)

Figure 1.27: Interior and exterior polygons - Situation 1.

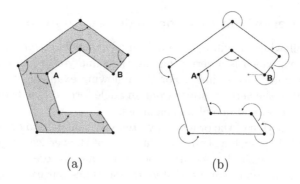

(a)   (b)

Figure 1.28: Interior and exterior polygons - Situation 2.

We will now present two methods to characterize interior and exterior polygons. The first one is classical and based on a geometrical calculation of the angles, whereas the second one is based on the vertex of the polygon having the smallest x-coordinate.

1.11.3.1 Angle-based method

For any polygon, the sum of its exterior angles is 360°. As we can see in Figure 1.29, if we imagine *walking* or, in other terms, *visiting* all the way around the outside of a polygon, we make one full turn:

$$s_{ext} = a + b + c + d + e = 360 \tag{1.3}$$

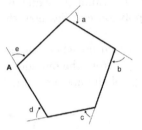

Figure 1.29: Sum of the exterior angles of a polygon.

Figure 1.30 illustrates for three examples that the sum s of interior angles of any polygon with n sides is given by:

$$s_{int} = (n - 2) \times 180 \tag{1.4}$$

This is due to the fact that each polygon with n sides can be divided into $n - 2$ triangles, and we know that the sum of interior angles of a triangle is equal to 180°.

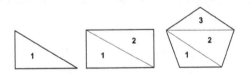

Figure 1.30: Sum of the interior angles of a polygon.

Based on Equation (1.3), we can write Equation (1.4) as follows:

$$s_{int} = 180 \times n - 360 \tag{1.5}$$

Note that these two properties are not verified directly in the case of α-polygons. They depend on the number of the α-loops and their positions.

Using Equation (1.4), we can therefore distinguish interior from exterior polygons where the angles are visited following the polar order.

One of the advantages of this method is that it can work for some pseudo-polygons (cf. Definition 39, Section 1.10.2), for instance, for an interior polygon containing some subpolygons that are connected to some of its vertices with a chain, as shown by Figure 1.31(a). The overall pseudo-polygon has 17 vertices and, based on Equation (1.4), the sum of its interior angles should be equal to $(17-2) \times 180° = 2700°$. However, if we calculate the real sum of its angles, we will find $3060°$ which is equal to $(19-2) \times 180°$. It is as if the polygon had 19 vertices instead of 17, which is correct since vertices A and B are visited twice. Figure 1.31(b) shows the polygon slightly modified, based on Definition 37, where we can assume that vertices A and B are doubled. Thus, we obtain a torn polygon with the same number of angles as the one in (a) but with 19 vertices. This second polygon clarifies how the vertices are visited in the pseudo-polygon of (a).

Figure 1.32 shows the situation where the second triangular polygon is connected to the first polygon from the outside and which is not visited if we start from the interior angle of the first polygon $P1$.

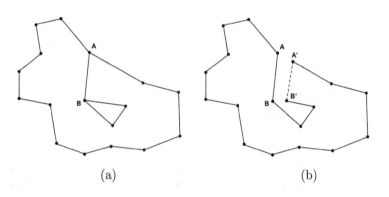

(a)   (b)

Figure 1.31: Repeatedly visiting some vertices of an interior polygon (1).

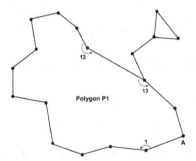

Figure 1.32: Repeatedly visiting some vertices of an interior polygon (2).

We can conclude from this situation that Equation (1.4) can be generalized by changing the definition of n which in reality represents the number of times that all the vertices or angles of a polygon have been visited.

There exist some specific structures that can cause confusion when visiting the vertices or angles of a polygon. These structures are those containing edges that intersect other edges of the same polygon, that we call a *tackled polygon* (cf. Definition 46). Different intersections can lead to different situations. We can give a list of the main structures like, for example, the *simple intersection* graph, *anchor* and *pseudo-anchor* graph, *boat* and *pseudo-boat* graph, etc. Figure 1.33 shows examples of such graphs. These specific structures look like polygons but integrating them into a polygon can cause two kinds of problems. The first is the non-verification of Equation (1.4), and the second is the visit of the interior as well as the exterior polygons before reaching the first visited angle.

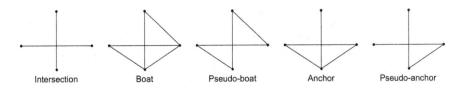

Figure 1.33: Specific structures.

Figure 1.34 shows another example that we call a β-*polygon*, in which an edge is replaced by a *boat graph* induced by the vertices A,B,C,D. This β-polygon has 17 vertices and the first problem posed by this example is its form. Is it a polygon? It is clear that it is not a polygon since vertex C does not belong to the polygon containing the edge $\{A, D\}$. Is it a pseudo-polygon? It is clear that it is not a pseudo-polygon either since the edge $\{A, D\}$ splits the overall polygon into two adjacent polygons. Whereas Equation (1.4) is not verified, we find that the sum of the angles of this β-polygon is equal to $(17 - 6) \times 180°$. Studying in detail this kind of graphs can be the subject of future research.

Figure 1.35 shows another example of a tackled polygon with a pseudo-anchor graph based intersection, containing two polygons separated by a chain of two edges, where one of them intersects with the first polygon. In the first case (see Figures 1.35(a) and (b)), the second polygon is outside the first, and in the second case (see Figures 1.35(c) and (d)) the second polygon is inside the first polygon. Depending on the starting vertex in the first polygon, these cases may result in two situations. The first is visiting only the first polygon (see Figures 1.35(b) and (d)), and the second is visiting all the vertices normally (see Figures 1.35(a) and (c)).

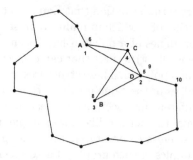

Figure 1.34: A boat graph as induced in a polygon (β-polygon).

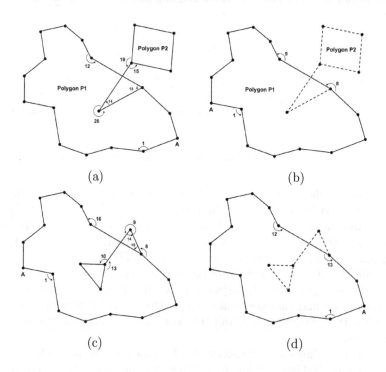

Figure 1.35: Visiting vertices of a tackled polygon - Example 1.

The polygon of Figure 1.36 shows an example of a polygon containing an anchor graph. The specificity of this polygon is that from the starting vertex A, we will visit all its interior and exterior angles before reaching the first visited angle a second time.

The last example of Figure 1.37 will help us to describe with accuracy when the situation of the previous polygon will arise. If we start from vertex A as in the previous example, we will visit all the interior and exterior angles

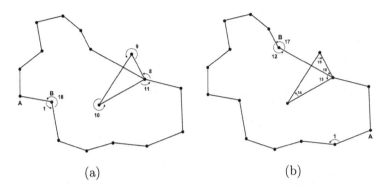

(a)   (b)

Figure 1.36: Visiting vertices of a tackled polygon - Example 2.

of this pseudo-polygon with intersection. We can conclude from this polygon that if a polygon has a chain connecting two of its vertices, then visiting its angles until reaching the first visited angle will lead to a visit of all its interior and exterior angles.

We finish by presenting some more situations that can be confusing. The first one may be called the *degree-1 starting vertex* and is illustrated by Figures 1.38(a) and (b). If we start the visit from B having the smallest x-coordinate and one neighbor, vertex B will be visited a second time with an angle α equal to 360°. This situation cannot exist in the case of interior polygons, as shown by the same figure, where after visiting the angle β, it is impossible to reach vertices B or D. Therefore, we conclude that if the vertex with the smallest x-coordinate of the polygon is visited with an angle equal to 360°, then this polygon is an exterior polygon.

Another confusing situation is shown by Figure 1.39, where the vertex with smallest x-coordinate of an interior polygon has many neighbors. This will lead to visiting many angles and then returning to that vertex. In this figure, vertex B is visited 3 times with the angles α, β and γ. Since the main angle $\widehat{ABC}$ is an interior one, it is smaller than 180°. Since any subangle of $\widehat{ABC}$ will be smaller, each of the angles α, β and γ is smaller than $\widehat{ABC}$ and therefore, smaller than 180°. We can conclude that it is sufficient to take into account the last visited angle γ to determine whether the obtained polygon is interior or exterior.

1.11.3.2 Minimum x-coordinate based method

For convenience and throughout the following, we will call the *global minimum* the vertex with the smallest x-coordinate.

A second possibility to characterize interior and exterior polygons is based on the vertex having the minimum x-coordinate in the obtained polygon hull. We just recall that a local minimum with respect to the x-coordinate is a vertex

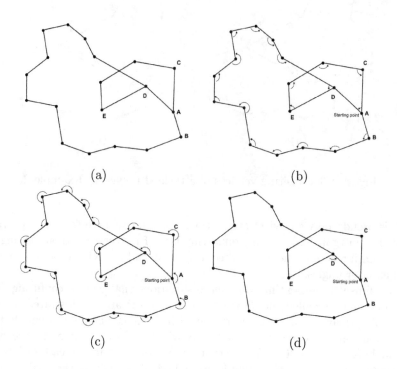

Figure 1.37: Visiting vertices of a tackled polygon - Example 3.

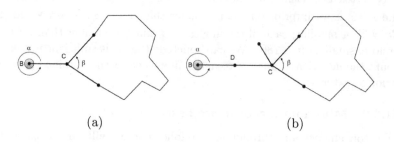

Figure 1.38: A starting vertex of degree 1.

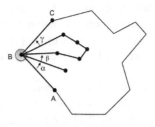

Figure 1.39: A starting vertex of degree greater than 1.

having no neighbor with an x-coordinate smaller than its own. Figure 1.40 shows such a local minimum B, where in (a) and (b), the y-coordinate of A is smaller than that of C, and in (c), the y-coordinate of A is greater than that of C. The global minimum has the same property since it is always a local minimum.

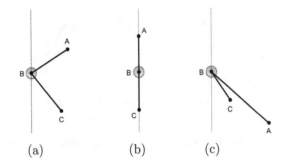

(a) (b) (c)

Figure 1.40: Possibilities for local minima with respect to the x-coordinate.

Another characteristic of local and global minima is related to the order of their visit with respect to their neighbors. Figures 1.41 and 1.42 show examples of a local/global minimum visited differently. In Figure 1.41(a), vertex B is visited after A and forming a polar angle greater or equal to 180°. However, in Figure 1.42(a), vertex B is visited after C with a polar angle smaller or equal to 180°.

If we look at the cases of Figures 1.41(b) and 1.42(b), where the visited angle is equal to 180°, we observe that the order of visiting a local minimum *can* depend on the y-coordinates of its neighbors, i.e., if the y-coordinate of A is smaller than that of C, then visiting B after A will lead to an angle $\overset{\frown}{ABC}$ greater or equal to 180°, and visiting B after C will lead to an angle $\overset{\frown}{CBA}$ smaller or equal to 180°. However, if we look at the cases of Figures 1.41(c) and 1.42(c), we obtain the same conditions for the angles as in (a) but with a y-coordinate of A greater than that of C. Therefore, the order of visiting a local minimum vertex depends on the value of the angle formed with its neighbors in case where this value is greater or smaller than 180°, and only if this value is equal to 180°, the order depends on the values of the neighbors' y-coordinates.

Altogether, we conclude that if the visited angle formed by a local minimum with its neighbors is greater than 180°, then this angle is visited from its exterior part, which means to the left of the local minimum and if this angle is smaller than 180°, then it is visited from its interior part, which means to the right of the local minimum. Finally, if this angle is equal to 180°, then we look at the y-coordinates of the neighbors of the local minimum. If the previous neighbor has a y-coordinate smaller than that of the subsequent neighbor, then the angle is visited from its exterior part, otherwise, it is visited from its interior part.

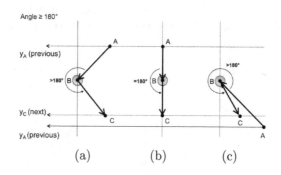

Figure 1.41: Possibilities of visiting a local minimum - Situation 1.

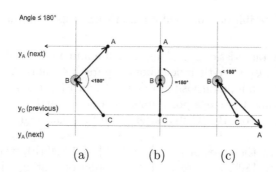

Figure 1.42: Possibilities of visiting a local minimum - Situation 2.

The main advantage of this characteristic is that if vertex B is the global minimum of the found polygon, the value of the angle formed with its previous and subsequent vertices can help to know whether this polygon is interior or exterior. Note that, if the vertex B is visited many times, then we consider only the last visited angle. Figure 1.43 shows how to differentiate an interior polygon from an exterior polygon after visiting the vertices in polar order. In Figure 1.43(a), the visit is started from the global minimum B. Then, once the stop condition is verified, i.e., the first angle or the vertex C is visited a second time from vertex B, the last angle obtained for B is greater than 180°. Therefore, the obtained polygon is an exterior polygon. However, in the case of Figure 1.43(b), the visit is started from vertex S, and once the stop condition verified, i.e., vertex T is visited a second time from vertex S, the last angle obtained for B is smaller than 180°. Therefore, the obtained polygon is an interior polygon.

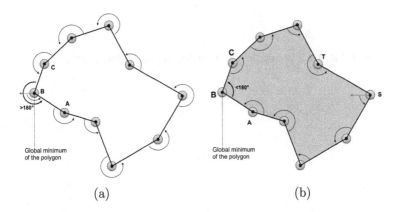

(a)   (b)

Figure 1.43: An exterior and an interior polygon.

If we use the same principle in the case of graphs instead of just polygons, we can obtain the same results, as shown by Figure 1.44 where for a visit from the starting vertex S, the last angle obtained for the global minimum B is smaller than 180°. Therefore, the obtained polygon is an interior polygon. Detecting interior polygons in graphs can help to detect gaps or voids in real network applications. This will be discussed in detail in Section 6.3.

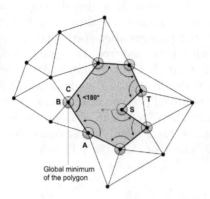

Figure 1.44: An interior polygon within a connected Euclidean graph.

Chapter 2

Hulls of point sets and graphs

Circumscribing a set of points in the plane or covering the area created this way may be of interest for several applications. We might need to construct a boundary for various purposes: to recognize the area under surveillance in a military zone, to circumscribe an infected region in the medical field, to identify a fire zone, etc. In this chapter, we put forward some definitions that we deem necessary for the understanding of the subsequent chapters. The 'envelope' (or 'hull'), for instance, is a concept commonly used for the description of a set of disconnected points.

In our work, the situation is different: we are given a connected graph and we are looking for a closed polygon defined by a subset of vertices and edges of the graph, which contains all the remaining vertices (and sometimes also the remaining edges) in its interior. Notions like 'convex, concave or polygon hull' that we are going to review now are already used in various contexts of computational geometry. What makes our work original and new is the condition on the polygon to exclusively use edges of the given graph to connect pairs of consecutive vertices.

2.1 Convex hull

2.1.1 Definitions and properties

Definition 47. We say that $C \subset \mathbb{R}^n$ is a *convex set* if and only if for any pair of points in C the whole line segment between these points is contained in C, or more formally, if and only if

$$\forall x, y \in C, \ and \ \forall \lambda \in [0,1], \lambda x + (1 - \lambda)y \in C \tag{2.1}$$

Geometrically, C is a convex set if and only if the whole connecting line segment between any two points of C is also part of C, i.e.,

$$C \text{ is convex iff } [x, y] \subseteq C, \forall x, y \in C \tag{2.2}$$

Figure 2.1 illustrates this property.

Definition 48. A set C^* in the Euclidean space $\mathbb{R}^n$ is called *star-shaped* if there

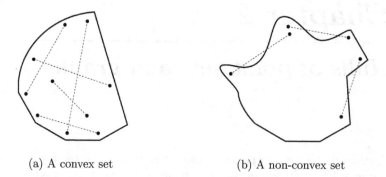

(a) A convex set   (b) A non-convex set

Figure 2.1: A convex and a non-convex set.

exists a point $x_0 \in C^*$ such that for every $x \in C^*$ the line segment $[x_0, x]$ is included in S^*.

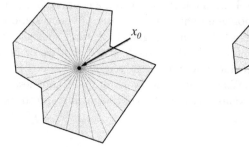
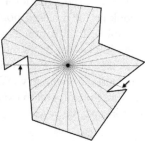

(a) A star-shaped set   (b) A non-star-shaped set

Figure 2.2: Examples of a star-shaped and a non-star-shaped set in $\mathbb{R}^2$.

The following are fundamental properties of convex and star-shaped sets:

1. The intersection of two convex sets is convex.

2. The intersection of two star-shaped sets is star-shaped.

3. The intersection of a convex set with a star-shaped set is star-shaped.

Definition 49. The *convex hull* $conv(S)$ of a set $S \subset \mathbb{R}^n$ is the intersection of all convex supersets of S, or in other words, the smallest convex set containing S.

The convex hull of a finite set $S \subset \mathbb{R}^n$ is also called a *convex polytope*.

Definition 50. A *vertex* of $conv(S)$ is a point p of S for which $p \notin conv(S \setminus p)$. Such a point is also called an *extreme point* of $conv(S)$.

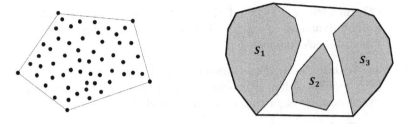

Figure 2.3: Convex hulls of a point set and a family of such sets.

There exists another definition of an extreme point that we can formulate as follows:

Definition 51. A point p of a convex set S is an *extreme point* of S if there are no two points a, b of S such that p lies on the open line segment $]a, b[$.

Proposition 3. [60] p is an extreme point of a polytope P if no triangle of vertices, different from p, constitutes a triangle containing p.

Constructing the convex hull of a finite point set S is a fundamental problem in computational geometry.

Optimizing a linear function f over such a set S in $\mathbb{R}^n$, for instance, reduces to optimizing f over the convex hull of the elements of S, a polytope in $\mathbb{R}^n$, the extreme points of which always contain an optimal solution.

This idea of limiting the effort to a reduced number of solutions, applied to so-called linear relaxations, has allowed the efficient solution of number of NP-hard combinatorial optimization problems.

The following is a list of further properties of convex sets.

Proposition 4. [50] A set $S \subseteq \mathbb{R}^n$ is convex if and only if it contains all convex combinations of elements of S, i.e., points of the form

$$\sum_{i=1}^{k} \lambda_i x_i \text{ with } x_i \in S, \lambda_i \geq 0 \text{ for } i = 1, .., k \text{ and } \sum_{i=1}^{k} \lambda_i = 1 \qquad (2.3)$$

Proposition 5. A convex polytope in $\mathbb{R}^n$ is the convex hull of its extreme points.

Theorem 7 (Radon 1921). [98] Any set $S \subset \mathbb{R}^n$ of $n + 2$ points can be partitioned into two disjoint subsets S_1 and S_2 such that $conv(S_1) \cap conv(S_2) \neq \emptyset$.

Theorem 8. [39] The convex hull of a set of n points in the plane can be computed in $O(n \log n)$ time.

Theorem 9 (Carathéodory). [31] For any $P \subset \mathbb{R}^n$ and $p \in conv(P)$ there exist $k \leqq n + 1$ points $p_1, p_2, ..., p_k \in P$ such that $p \in conv(p_1, p_2, ..., p_k)$.

2.1.2 Examples

The concept of a convex hull is applied in various fields. We can cite:

Fluid Dynamics LeFloch and Mercier [79] used the convex hull properties in fluid dynamics to extract dissipative entropy solutions after a shock-formation environment.

Mobile Ad-hoc Networks Kundu et al. [74] applied the convex hull to deal with mobile ad-hoc networks and proposed a new on-demand power saving routing algorithm for such networks.

Aircraft Trajectory Planning Pierre et al. [96] suggested an approach in which paths in a two-dimensional space are designed with the help of convex hull generation to deal with aircraft trajectory planning problems.

Routing and Geo-casting Stojmenovic et al. [107] proposed an algorithm for convex hull construction based on a unified framework for both routing and geo-casting problems, in which a message is forwarded to exactly those neighbors that may be best choices for a possible position of destination.

Geometric Matching Scheduling Hershberger et al. [59] solve, by using the convex hull concept, the following two optimization problems:

1. 'Matching': given red points and blue points on the plane, find a matching of red and blue points, by line segments, in which no two edges cross.

2. 'Scheduling': given jobs with due dates, linear penalties for late completion, and a single machine on which to process the jobs, find a schedule of executing the jobs that minimizes the maximum penalty.

Process Operation and Scheduling Pham et al. [95] recommended a new discretization approach allowing to linearize the bilinear terms (flow variables multiplied by quality variables), where the quality variables are discretized. Implicit enumeration of the discretization is used to obtain a convex hull which restricts the size of the exploration space.

2.2   Affine hull

2.2.1   Definitions and properties

Definition 52. A non-empty subset F of $\mathbb{R}^n$ is called an *affine subspace* if:

$$\forall x, y \in F, \text{ and } \forall \lambda \in \mathbb{R}, \lambda x + (1 - \lambda)y \in F \qquad (2.4)$$

Proposition 6. Affine subspaces have the following properties:

1. If F_1 and F_2 are two affine subspaces of $\mathbb{R}^n$ and $\lambda_1, \lambda_2 \in \mathbb{R}$, then $\lambda_1 F_1 + \lambda_2 F_2$ is an affine subspace of $\mathbb{R}^n$.

2. If F_1 and F_2 are two affine subspaces of $\mathbb{R}^n$, then $F_1 \cup F_2$ is an affine subspace of $\mathbb{R}^n$.

3. If F is an affine subspace of $\mathbb{R}^n$ and f an affine application, i.e., $f(x) = Ax + b$ for $A \in \mathbb{R}^{m \times n}$ and $b \in \mathbb{R}^m$, then $f(F)$ is an affine subspace of $\mathbb{R}^m$.

Definition 53. Let F be an affine subspace of $\mathbb{R}^n$. The *vector space* associated with F is defined by:

$$lin(F) = \{v \in \mathbb{R}^n | \forall x \in F, x + v \in F\} \qquad (2.5)$$

Definition 54. Let $x_1, ..., x_k$ be a finite number of points of $\mathbb{R}^n$ and $\lambda_1, ..., \lambda_k$ real numbers such that:

$$\sum_{i=1}^{k} \lambda_i = 1 \qquad (2.6)$$

Then $x = \sum_{i=1}^{k} \lambda_i x_i$ is called an *affine combination* of $x_1, .., x_k$.

Definition 55. A subset F of $\mathbb{R}^n$ is an *affine subspace* if and only if it contains all affine combinations of points of F.

Definition 56. The *affine hull* of $S \subseteq \mathbb{R}^n$ is the set of all affine combinations of S or, equivalently, the smallest affine subspace containing S.

2.2.2 Examples

Affine hulls give less accurate approximations of the boundary of a region than convex hulls, and are not as widely used as the convex hull. Their fields of application include:

Combinatorial Optimization: In this field, a basic problem is to optimize a linear function over the polytope P given as the convex hull of a finite point set in $\mathbb{R}^n$ [102]. To efficiently solve such a problem by linear programming techniques, one is interested in describing P by a set of linear inequalities and equations. A minimal such equation system immediately provides a description of the affine hull of P, and its rank the dimensional gap of P within $\mathbb{R}^n$.

Visual Tracking: Wang et al. [119] and Zhang et al. [120] represent the target candidates by affine combinations of a template set, which improves the capability in describing unseen target appearances.

Land Cover Classification: Huo et al. [64] present a new learning approach, called locally softened affine hull (LSAH), by introducing a new technique to solve the scarce sample problem in remote sensing.

Recognizing Objects: Wan et al. [116] consider the issue of identifying objects in egocentric videos where both red, green and blue (RGB) and depth data are available. The authors present a new dataset composed of RGB-Depth (RGB-D) egocentric videos capturing daily objects being manipulated during human activities. They use a kernel function that measures the similarity of two sets by the minimum distance between their sparse affine hulls.

2.3 α-shape

The human eye can easily distinguish the form generated by a set of points or a connected graph, and thus recognize its resemblance to existing objects. However, this exercise is not so easy to accomplish for a computer. The *α-shape* is one of the tools allowing him to execute such a task. α-shape is used as a generalization of the convex hull to captivate the shape of a set of points in the plane or in space. α-shape can be derived from Delaunay's triangulation, and it offers a concrete definition of a shape to represent the structure of a set of points [129]. In the following, we briefly describe this notion and give some properties as well as some examples.

In 1983, Edelsbrunner et al. [43] introduced the concept of α-shape associated with a set of points in the aim to generalize the idea of the convex hull. Their motivation was to find a boundary more refined than the one generated by a pivoting line segment to recognize the shape of objects. Replacing straight lines by discs of a fixed radius and considering the set of points of S not lying in any empty open disc gave rise to the notion of an α-hull of S, and substituting circular arcs by edges (imagining locally the shape of S on the side of the edge opposite to the center of the disc) finally led to the α-shape of S.

2.3.1   Definition and properties

Following Edelsbrunner [44], the α-shape of S can be defined as follows:

Definition 57. Let S be a finite set of points in the plane and α a non-negative real number. An open disk is said to be empty if it contains no point of S. Connecting by an edge any two points of S lying on the bounding circle of an empty disk of radius α we obtain the α-shape of S.

See Figure 2.4 for an illustration.

One can show that for $\alpha = 0$, the α-shape of S reduces to the set S itself, and for $\alpha = +\infty$, the α-shape of S coincides with the convex hull of S. Moreover, there are close structural and algorithmic relationships between the α-shape of S and the associated Delaunay triangulation and Voronoï dia-

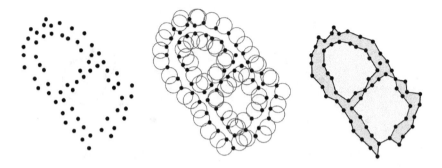

Figure 2.4: A set of points, its α-hull and its α-shape.

gram, both of which can also be viewed as projections of polytopes in $\mathbb{R}^3$ [15]. We also mention that three-dimensional α-shapes have been introduced in Edelsbrunner and Mücke [45].

Observation 2. If $\alpha_1 \leq \alpha_2$, then the α_1-hull of a set of points is contained in its α_2-hull.

Definition 58. A point p in a set S is called an α-*extreme point* in S if there exists a closed disc of radius α, such that p is on its border containing all the points of S.

Definition 59. A k-*simplex* Δ_T is a polytope of dimension k which is the convex hull of T and which has $k+1$ extreme points.

Example 10. For example,

- a 0-simplex is a single point,

- a 1-simplex is a line segment,

- a 2-simplex is a triangle,

- a 3-simplex is a tetrahedron.

Definition 60. An α-ball b is an open ball with radius α, and ∂b denotes the surface of the resulting sphere.

Definition 61. A k-simplex Δ_T is said to be α-exposed if there exists an empty α-ball b with $T = \partial b \cap S$.

Definition 62. The *boundary* ∂S_α of the α-shape of S consists of all α-exposed k-simplices of S. More formally,

$$\partial S_\alpha = \{\Delta_T | T \subset S, |T| \leq d \text{ and } \Delta_T \ \alpha\text{-exposed}\} \qquad (2.7)$$

Observation 3. The boundary ∂S_α of the α-shape is not necessarily the border (frontier) of the set S.

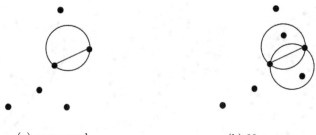

(a) α-exposed   (b) Not α-exposed

Figure 2.5: Examples of α-exposed and non-α-exposed simplices.

From Equation (2.7), we can obtain the following relations:

$$\lim_{\alpha \longrightarrow 0} S_\alpha = S \tag{2.8}$$

$$\lim_{\alpha \longrightarrow +\infty} S_\alpha = conv(S) \tag{2.9}$$

2.3.2 α-complex

Definition 63. Let $\Delta_T = conv(T)$ be a $k-$simplex, with $0 \leq k \leq d$ and spanned by some vertices of S. Then we let

- σ_T denote the radius of the circumsphere of Δ_T,
- μ_T denote the center of the circumsphere of Δ_T,

where the circumsphere is a circle in case that $d = 2$.

Definition 64. For a given set of points $S \subset \mathbb{R}^d$ and $0 \leq \alpha \leq \infty$, the *$\alpha$-complex* $C_\alpha(S)$ of S is the simplicial subcomplex of $DT(S)$ obtained as follows:

(C1) $\sigma_T < \alpha$ and the σ_T-ball located at μ_T is empty, or

(C2) Δ_T is a face of another simplex in $C_\alpha(S)$.

Proposition 7. [48] Let $\Delta_T \in \partial S_\alpha(S)$ be any simplex. Then $\Delta_T \in C_\alpha(S)$.

2.3.3   Relation between α-shape and Delaunay triangulation

For every simplex in the Delaunay triangulation, it is sufficient to note some intervals of α which define the status of that simplex. Then the entire class of shapes along with the complete history of the situation of each simplex can be reproduced by an annotated variant of the Delaunay triangulation. Once this annotated Delaunay triangulation is provided, the α-shape corresponding to any particular value of α can be dropped quickly and easily. It is possible to enable an interactive user to control the variable α and to visualize the resulting shapes from the convex hull to the original, disconnected set of points.

2.3.4 Examples

Beyond the straightforward applications in the fields of Pattern Recognition, Shape Reconstruction or Molecular Biology, the concept of α-shape has been used in Medical Image Analysis, Bioinformatics and Network Design. We can cite:

Boundary Detection in Wireless Sensor Networks: Fayed and Mouftah [47] have developed an algorithm to recognize nodes and edges along network boundaries. They addressed the problem of edge discovery using the α-shape technique.

Sampling and Reconstructing Manifolds: In [11], the authors formalized the problem of reconstruction of a form and gave sufficient conditions to reconstruct an object using α-shapes. They also discuss some practical considerations in reconstructing solids.

Protein-DNA Interactions: The protein-DNA interactions form a set of chemical bonds joining a protein to a DNA molecule. They occur in particular to regulate the biological functions of DNA. α-shape and Delaunay triangulation are powerful tools to represent protein structures and have advantages in characterizing the surface curvature and atom contacts [129].

More applications of the α-shape technique can be found in [45, 51, 89], and references therein.

2.4 Boundaries of graphs

As we have shown in the previous sections, there are different ways to attribute a shape to a given set of points. Of particular interest could be the notion of a boundary, a concept which so far has not been studied extensively in graph theory. The recent development of sensor technology has given the possibility to address new and important applications: securing the frontiers of a country, a dangerous region or a military territory by means of Wireless Sensor Networks, for instance, immediately leads to the concept of the boundary of an associated Euclidean graph whose nodes are given by the sensors (usually deployed in the plane) and whose edges represent the ability of a pair of sensors to directly communicate with each other.

In this section, we present different possibilities of defining the boundary of a given Euclidean graph, the main objective being the development of centralized and distributed versions of boundary detection algorithms as recently published in [18, 28, 24, 76, 100].

The different types of boundaries will be dealt with for our two types of graphs: plane Euclidean graphs and general Euclidean graphs. We prefer to distinguish between these two types because the boundary of a graph with edge-crossings may take different forms according to the application. Also observe that unlike non-planar graphs, a Euclidean graph with edge-crossings may not contain any subdivision of $K_{3,3}$ or K_5 as a partial subgraph. Figure 2.6(a) shows an example of a boundary of a Euclidean graph and Figure 2.6(b) shows its formed shape.

Definition 65. The *boundary* of a Euclidean graph $G = (V, E)$ is given by a cycle of G describing the outer face of G, i.e., circumscribing all vertices of G. A *boundary vertex* of G is a vertex whose deletion modifies the area formed by G.

Figure 2.6(c) shows an example where the shape of Figure 2.6(b) is modified after the deletion of a boundary vertex. However, the shape of Figure 2.6(d) remains the same as that of Figure 2.6(b) after the deletion of a non-boundary vertex.

2.4.1   Polygon hull of plane Euclidean graphs

We can formulate the *polygon hull* problem for Euclidean graphs as follows:

Definition 66. Given a Euclidean graph $G = (V, E)$ whose vertices are randomly deployed in the plane, the *polygon hull* problem for G is to find a cycle in G circumscribing all vertices of G.

2.4.2   Properties

Property 2. The polygon hull of a plane Euclidean graph describes its shape.

Proposition 8. A plane Euclidean graph has a unique polygon hull.

Proof. Follows from Property 2, since the shape of a plane Euclidean graph is unique.   □

The next subsection is about general Euclidean graphs, which are graphs that may contain edge-crossings. As already mentioned, these graphs are not necessarily non-planar.

2.4.3   Polygon hull of general Euclidean graphs

To our knowledge, the polygon hull problem for general Euclidean graphs has not been studied previously. In this textbook, we try to make a contribu-

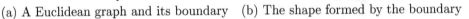

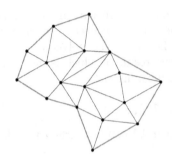

(a) A Euclidean graph and its boundary

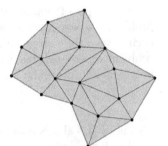

(b) The shape formed by the boundary

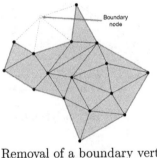

(c) Removal of a boundary vertex
modifies the shape

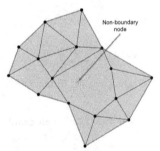

(d) Removal of a non-boundary vertex
leaves the shape unchanged

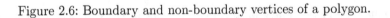

Figure 2.6: Boundary and non-boundary vertices of a polygon.

tion in this direction. We suggest three definitions of this problem; it will be up to the reader to decide which interpretation best suits his application.

In the following, we use these two notations:

- We denote a polygon by $P(V)$, where $V = \{v_1, v_2, ..., v_m\}$ is the set of its *vertices* and $\{\{v_1, v_2\}, \{v_2, v_3\}, ..., \{v_m, v_1\}\}$ the set of its *edges*.

- $\widehat{P}(V)$ is the area occupied by the polygon $P(V)$.

2.4.3.1 *A*-polygon hull

This type of boundary is often encountered in surveillance or security problems, where one wants to determine a polygon over a set of vertices in such a way that all remaining vertices lie inside the polygon. We call this problem the '*A-polygon hull*' problem, the result of which is described by a polygon circumscribing all vertices of V, meaning that:

$$\mathbb{V}_a \subseteq V \text{ and } v_i \in \widehat{P}(\mathbb{V}_a), \forall v_i \in V \tag{2.10}$$

Figure 2.7 shows an example of a general Euclidean graph together with an A-polygon hull.

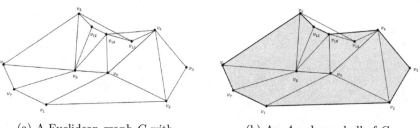

(a) A Euclidean graph G with edge-crossings

(b) An A-polygon hull of G

Figure 2.7: A general Euclidean graph and an associated A-polygon hull.

2.4.3.2 *B*-polygon hull

This type of boundary problem can be met in the case where we want to find a minimal boundary of the given network, i.e., a polygon circumscribing all vertices of the network and its own vertices being of minimum cardinality. We call this problem the *B-polygon hull* problem, which can be formulated as follows:

$$\begin{cases} \underset{v \in V}{\text{minimize}} & |\mathbb{V}_b| \\ \text{subject to} & v_i \in \widehat{P}(\mathbb{V}_b), \forall v_i \in V \\ & \mathbb{V}_b \subseteq V \end{cases} \tag{2.11}$$

Figure 2.8 shows a B-polygon hull of the graph depicted in Figure 2.7(a).

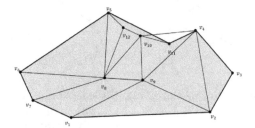

Figure 2.8: A general Euclidean graph and an associated B-polygon hull.

Observation 4. A B-polygon hull is always an A-polygon hull, but the converse does not hold.

2.4.3.3 C-polygon hull

The area covered by a Euclidean graph in the plane gives rise to several applications as for instance the coverage area problem in Wireless Sensor Networks [63]. This third type of polygon hull is given by '*the boundary delimiting the maximal area occupied by the graph*'. It will be called a *C-polygon hull*, and the related optimization problem can be formulated as follows:

$$\begin{cases} \underset{v \in V}{\text{minimize}} & f_1(v) = |\mathbb{V}_c| \\ \underset{v \in V}{\text{maximize}} & f_2(v) = \widehat{P}(\mathbb{V}_c) \\ \text{subject to} & v_i \in \widehat{P}(\mathbb{V}_c), \forall v_i \in V \\ & \mathbb{V}_c \subseteq V \end{cases} \quad (2.12)$$

where $\widehat{P}(\mathbb{V}_c)$ can be calculated by Equation (1.2).

Figure 2.9 shows an example of a boundary delimiting the maximal area occupied by a general Euclidean graph.

The C-polygon hulls depicted in Figure 2.9(a) and Figure 2.9(b) are feasible for Problem (2.12) since they include all the graph's vertices. They are, however, not necessarily optimal with respect to the two objective functions. Such solutions are also called *efficient*, because an improvement of one objective can only be obtained by a deterioration of the other[1].

The solution shown in Figure 2.9(c) optimizes both objectives, but is unfortunately not feasible due to the intersection between the edges $\{v_{10}, v_4\}$ and $\{v_5, v_{11}\}$ as given by the vertex v_f which is not a vertex of the initial graph. This intersection is represented by a fictive vertex v_f in Figure 2.9(d), which represents an ideal solution to Problem (2.12), but not a feasible one because $v_f \notin V$.

---

[1]According to the definition of efficiency given by V. Pareto [92].

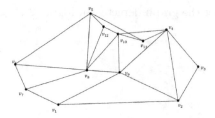

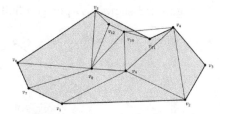

(a) A feasible solution to Problem (2.12)

(b) A feasible solution to Problem (2.12) with an improvement of the first objective (cardinality of $\mathbb{V}_c$) obtained by a deterioration of the second one (area)

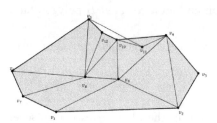

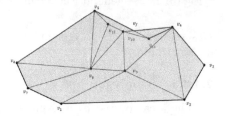

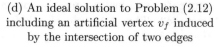

(c) Another feasible solution to Problem (2.12) with an improvement of the second objective obtained by a deterioration of the first one

(d) An ideal solution to Problem (2.12) including an artificial vertex v_f induced by the intersection of two edges

Figure 2.9: A general Euclidean graph and an associated C-polygon hull.

These cases will be treated in detail in Chapter 3.

Observation 5. All three problems are equivalent in the case of plane Euclidean graphs.

Chapter 3

Centralized algorithms for boundary detection

During the 1970s, an important number of methods has been proposed for finding the convex hull of a set of points. In 1972, Graham was the first to propose a technique based on scanning points from right to left. One year after, Jarvis proposed a new technique based on polar angles, and then a number of other methods have been published including Andrew, Kallay, Chan, etc. Some of these algorithms are presented in Section 3.1. If the problem of finding the convex hull attracted the curiosity of several researchers, unfortunately, this is not the case for concave and polygon hulls. In Section 3.2, we present some of the existing algorithms for the construction of a concave hull. Finally, Sections 3.3 and 3.4 address some new algorithms for finding a polygon hull in connected Euclidean graphs like the LPCN algorithm. We present several versions of LPCN that can be applied to different types of the considered graphs. We conclude with a new algorithm, called RR-LPCN, which has been designed for the same purpose, but for which it is not necessary to specify the starting node.

3.1 Finding the convex hull of a set of points in the plane

The computation of convex hulls (often called convex envelopes) is one of the most famous and well-studied problems of computational geometry, which explains the large quantity of procedures developed in this context. The purpose of this chapter is to present some of the most popular methods for the calculation of envelopes, and above all to give an overview of some algorithmic paradigms.

The problem we are going to study arises as follows: given a set S of n points in the plane, find those that form a convex polygon containing all remaining points.

An intuitive image is that of wrapping planted nails with an elastic, the result of which will form the convex envelope of these nails, as shown in Figure 3.1.

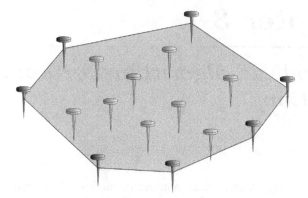

Figure 3.1: Wrapping nails with an elastic.

3.1.1   Jarvis' algorithm

The first algorithm we will examine was designed by R. A. Jarvis [66] in 1973. He was inspired by the gift wrapping procedure to find a convex envelope of a set S of points in the plane. Its principle is to start with a point located at an extreme end and then to select the point that forms the minimum polar angle with respect to the current point. At the end, two objectives must be verified:

1. all points of the set are covered,

2. a convex polygon is formed.

The algorithm starts by selecting a point P_c with minimum x-coordinate, i.e., a point located at the extreme left. The set $\mathbb{B}$, which will represent the convex hull of S, is initialized to $\{P_c\}$, and a fictitious point P_p is chosen to the left of P_c. After that, we look for a point P_k belonging to S and having the minimum polar angle formed by the segments $[P_0, P_k]$ and $[P_p, P_c]$. Then the variables are updated: we add point P_k to $\mathbb{B}$, and we replace P_p by P_c, P_c by P_k.

Jarvis' algorithm is described by Algorithm 1.

Algorithm 1: Jarvis

Data: S, a set of points in the plane
Result: $\mathbb{B}$, the set of points forming the convex hull of S
1 $P_c \leftarrow$ a point with minimum x-coordinate;
2 $\mathbb{B} \leftarrow \{P_c\}$;
3 $P_{first} \leftarrow P_c$;
4 $P_p \leftarrow$ a fictitious point situated to the left of P_{first};
5 **repeat**
6 $\quad P_k = \underset{P_j \in V}{\mathrm{argmin}}\{\varphi(P_p, P_c, P_j)\}$;
7 $\quad \mathbb{B} \leftarrow \mathbb{B} \cup \{P_k\}$;
8 $\quad P_p \leftarrow P_c$;
9 $\quad P_c \leftarrow P_k$
10 **until** $P_k = P_{first}$;

Figure 3.2 shows an example of the execution of Jarvis' algorithm.

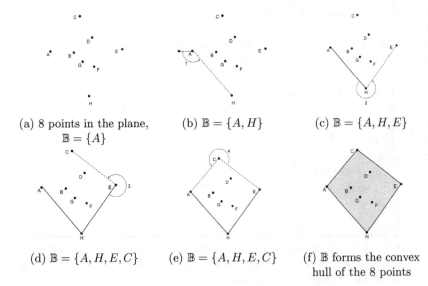

(a) 8 points in the plane, $\mathbb{B} = \{A\}$

(b) $\mathbb{B} = \{A, H\}$

(c) $\mathbb{B} = \{A, H, E\}$

(d) $\mathbb{B} = \{A, H, E, C\}$

(e) $\mathbb{B} = \{A, H, E, C\}$

(f) $\mathbb{B}$ forms the convex hull of the 8 points

Figure 3.2: An example for Jarvis' algorithm.

The complexity of Jarvis' algorithm is $O(nh)$, where h is the number of vertices on the resulting convex envelope, and n the number of points in S. If the number of vertices lying on the convex envelope is high, the complexity approaches $O(n^2)$. In [1], Akl proposed an improvement of Jarvis' algorithm by dividing the set of all points into two subsets and modifying the angle orientation within each set.

3.1.2   Graham's algorithm

This algorithm, also called *Graham scan algorithm*, owes its name to Ronald Graham who published the original algorithm in 1972 [54]. Graham is considered a pioneer in computational geometry.

The principle is to start with a point P_1 "at the bottom of S", i.e., a point with minimum y-coordinate. Then a sorting is performed on all other points $P_i, i \neq 1$, according to the order of the angles formed by the segment $[P_1, P_i]$ with the horizontal axis. By using the direction of the angle formed after every two successive segments, a decision is made on P_i whether or not to belong to the convex envelope. The algorithm stops when the starting point is reached for the second time.

Graham's algorithm is described by Algorithm 2.

Algorithm 2: Graham

   Data: A table containing the points S
   Result: $\mathbb{B}$, the set of points forming the convex hull of S
1 Find P_1, a point with minimum y-coordinate;
2 $\mathbb{B} = \{P_1\}$;
3 Sort the points of $S \backslash \{P_1\}$ by increasing angle with respect to P_1;
4 Let S' be the resulting set of ordered points of S;
5 $j = 1; \mathbb{B} = \mathbb{B} \cup \{P_2\}$;
6 **for** $P_i \in S' \backslash \{P_1\}, i \geq 3$ **do**
7   **if** P_i *is located to the left of the line* $[P_j, P_{j+1}]$ **then**
8        $\mathbb{B} = \mathbb{B} \cup \{P_i\}$;
9        $j = j + 1$;
10   **else**
11       $\mathbb{B} = \mathbb{B} \backslash \{P_i\}$;
12       $j = j - 1$;
13   **end**
14 **end**

Figure 3.3 shows an example of the execution of Graham's algorithm.

The complexity of the sorting procedure is $O(n \log n)$, where n is the number of points in S. Selecting the points on the convex envelope can be done in $O(n)$.

Observation 6. Graham's algorithm was the first algorithm of optimal complexity to construct a convex envelope. There are several variants of this algorithm, we can cite [2, 3, 5, 6, 55, 73, 105, 112] but some of them are incorrect, either in their descriptions or their implementation [56].

3.1.3   The Quickhull algorithm

The Quickhull algorithm was discovered independently by W. Eddy [42] in 1977 and A. Bykat [30] in 1978. Inspired by the Quicksort algorithm [88], its principle is the use of the *divide and conquer* strategy.

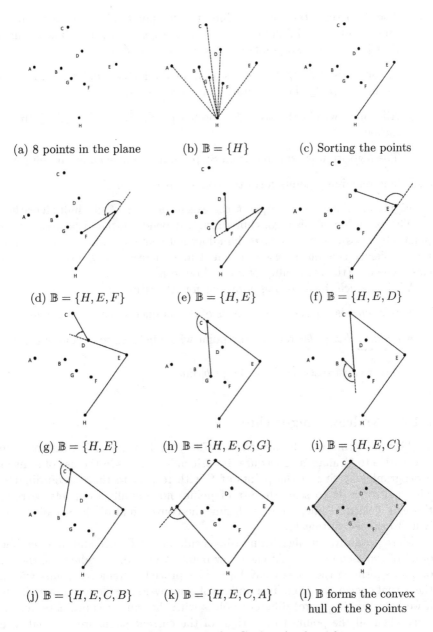

(a) 8 points in the plane (b) $\mathbb{B} = \{H\}$ (c) Sorting the points

(d) $\mathbb{B} = \{H, E, F\}$ (e) $\mathbb{B} = \{H, E\}$ (f) $\mathbb{B} = \{H, E, D\}$

(g) $\mathbb{B} = \{H, E\}$ (h) $\mathbb{B} = \{H, E, C, G\}$ (i) $\mathbb{B} = \{H, E, C\}$

(j) $\mathbb{B} = \{H, E, C, B\}$ (k) $\mathbb{B} = \{H, E, C, A\}$ (l) $\mathbb{B}$ forms the convex hull of the 8 points

Figure 3.3: An example for Graham's algorithm.

Let S be a set of points in the plane. The Quickhull algorithm aims to find the convex hull of S and can be summarized by the following steps:

1. The algorithm starts by searching for two points $\bar{P}, \underline{P}$ having the maximum and the minimum x-coordinate, respectively. The line segment that joins these two points divides S into S_1 and S_2.

2. In each subset S_1 and S_2, we search for the farthest point with respect to the line $[\bar{P}, \underline{P}]$. Let P_1 and P_2 be these points.

3. All points within the triangles formed by $P_1, \bar{P}, \underline{P}$ and $P_2, \bar{P}, \underline{P}$ are removed.

4. The algorithm is repeated from Step 2 until all assemblies are empty.

5. The remaining points form the convex envelope of S.

Figure 3.4 shows an example of the execution of the Quickhull algorithm.

The execution of this algorithm strongly depends on the disposition of the points. In cases where the convex hull does not contain many extreme points, its complexity is about $O(n \log n)$, but it may increase to $O(nh) \approx O(n^2)$ in cases where h (the cardinality of $\mathbb{B}$) is close to n.

We introduce the following notations for Algorithm 3:

- $\Delta(S, abc)$ is the set of the points of S contained in the triangle abc.

- $\underbrace{argmax}_{x \in X} d(x, [a, b])$ returns the point which belongs to X and which has the greatest distance from the $[a, b]$ line.

3.1.4   Andrew's algorithm

This algorithm was discovered in 1979 by Andrew [6]. It is considered an alternative to Graham's procedure. Its principle consists in the use of a linear lexicographic sorting of the points of S with respect to their x-coordinates. The main idea is to scan the set of points horizontally from left to right through a vertical line, while updating the convex hull of the set of points lying to the left of that line.

More precisely, the algorithm visits each point of P once and in ascending order of the x-coordinate. At each new point P_i visited, it updates $\mathbb{B}$, the set of points forming the convex envelope. This procedure removes points within the actual convex polygon. The update of the envelope is done symmetrically if we have two (or more) different points with the same x-coordinate, and it ends when all the points to the right of the current point are reduced to a single item.

Figure 3.5 shows an example of the execution of Andrew's algorithm. It is easy to observe that both the first and last point of this path belong to the convex envelope.

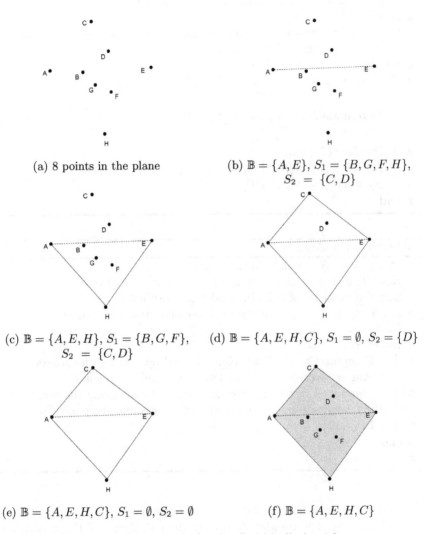

(a) 8 points in the plane

(b) $\mathbb{B} = \{A, E\}$, $S_1 = \{B, G, F, H\}$, $S_2 = \{C, D\}$

(c) $\mathbb{B} = \{A, E, H\}$, $S_1 = \{B, G, F\}$, $S_2 = \{C, D\}$

(d) $\mathbb{B} = \{A, E, H, C\}$, $S_1 = \emptyset$, $S_2 = \{D\}$

(e) $\mathbb{B} = \{A, E, H, C\}$, $S_1 = \emptyset$, $S_2 = \emptyset$

(f) $\mathbb{B} = \{A, E, H, C\}$

Figure 3.4: An example for the Quickhull algorithm.

Algorithm 3: Quickhull

Data: A set S of points in the plane

Result: $\mathbb{B}$ the set of points forming the convex hull of S

1 Let $\underline{P}$ be a point with minimum x-coordinate;

2 Let $\bar{P}$ be a point with maximum x-coordinate;

3 $\mathbb{B} = \{\bar{P}, \underline{P}\}$;

4 Let S_1, S_2 respectively, denote the upper and lower subsets resulting from the separation of S by $[\bar{P}, \underline{P}]$;

5 **while** $S_1 \neq \emptyset$ *or* $S_2 \neq \emptyset$ **do**

6   Determine $P_1 = \underbrace{argmax}_{P_i \in S_1} d(P_i, [\bar{P}, \underline{P}])$;

7   Determine $P_2 = \underbrace{argmax}_{P_i \in S_2} d(P_i, [\bar{P}, \underline{P}])$;

8     $\mathbb{B} = \mathbb{B} \cup \{P_1, P_2\}$;

9     $S_1 = S_1 \backslash \{P_1, \Delta(S, \underline{P}\bar{P}P_1)\}$;

10    $S_2 = S_2 \backslash \{P_2, \Delta(S, \underline{P}\bar{P}P_2)\}$;

11 **end**

Algorithm 4: Andrew

Data: A set S of points in the plane

Result: $\mathbb{B}$, the set of points forming the convex hull of S

1 Sort the elements of S by increasing x-coordinate;

2 Let $S' = (s_1, s_2, .., s_n)$ be the sequence of sorted points of S;

3 $i = 1$;

4 **while** $S' \neq \emptyset$ **do**

5   - Compute $\mathbb{C}_L^i$, the lower chain formed by deleting the points in the triangles given by the lower left points and the point P_i;

6   - Compute $\mathbb{C}_U^i$, the upper chain formed by removing the points in the triangles given by the upper left points and the point P_i;

7   - $i = i + 1$;

8 **end**

9 $\mathbb{B} = C_L \cup C_U$;

Andrew's algorithm is described by Algorithm 4.

If the points of S are already sorted, the complexity of this algorithm is $O(n)$. Otherwise, its complexity is $O(n \log n)$.

If more than two points have the same x-coordinate, we must consider the y-coordinate to sort these points. The lower (resp. upper) points are the points which have the coordinate $y_{P_i} < y_{P_0}$ (resp. $y_{P_i} \geq y_{P_0}$).

Observation 7. This algorithm is also called *"monotone chain algorithm"* because it builds a chain of points with increasing x-coordinates.

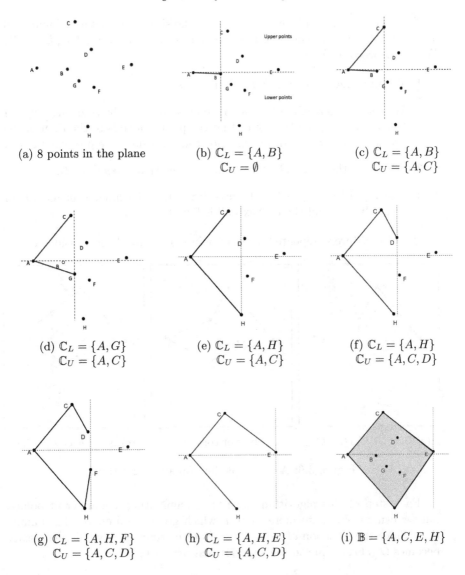

(a) 8 points in the plane

(b) $\mathbb{C}_L = \{A, B\}$
$\mathbb{C}_U = \emptyset$

(c) $\mathbb{C}_L = \{A, B\}$
$\mathbb{C}_U = \{A, C\}$

(d) $\mathbb{C}_L = \{A, G\}$
$\mathbb{C}_U = \{A, C\}$

(e) $\mathbb{C}_L = \{A, H\}$
$\mathbb{C}_U = \{A, C\}$

(f) $\mathbb{C}_L = \{A, H\}$
$\mathbb{C}_U = \{A, C, D\}$

(g) $\mathbb{C}_L = \{A, H, F\}$
$\mathbb{C}_U = \{A, C, D\}$

(h) $\mathbb{C}_L = \{A, H, E\}$
$\mathbb{C}_U = \{A, C, D\}$

(i) $\mathbb{B} = \{A, C, E, H\}$

Figure 3.5: An example for Andrew's algorithm.

3.1.5 Kallay's algorithm

Invented by Michael Kallay [70] in 1984, this procedure is considered a variant of Andrew's algorithm [6] and it is also called incremental algorithm. It consists in starting from three points randomly chosen followed by adding the points one by one and updating the convex hull at each step in order to ensure the accuracy.

Suppose that, at a given step, we have the situation shown in Figure 3.6 and we want to add the point E to the actual convex hull $\mathbb{B} = \{A, C, D, F, G\}$. We can proceed as follows:

1. We add edges that link E to D and to F.

2. We test the convexity of the area circumscribed by the resulting polygon (see Figure 3.6(b)): We notice that the point added is to the right of the previous edge. So we delete this point and consider the preceding point.

3. We delete the point D and we add an edge that links C to E.

4. The algorithm stops when it passes from the point having the maximum x-coordinate, and the convex hull is found.

This procedure is repeated until all points of S have been considered.

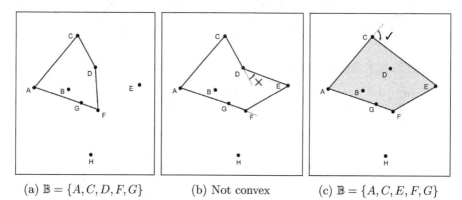

(a) $\mathbb{B} = \{A, C, D, F, G\}$ (b) Not convex   (c) $\mathbb{B} = \{A, C, E, F, G\}$

Figure 3.6: An example for Kallay's algorithm.

Each step of this algorithm consists in eliminating a number of points, but we cannot eliminate more than n, which gives the limit on the running time. Since the insertion of n points takes $O(n)$ time, the overall complexity becomes $O(n \log n)$ due to the superior cost of sorting.

3.1.6   Chan's algorithm

In 1996, Timothy Chan [32] proposed an algorithm which calculates the convex envelope of a set of points in two or three dimensions. It represents a combination of the algorithms of Jarvis and Graham, and is sensitive to its output, in particular to the cardinality of the points forming the convex envelope.

Chan's algorithm begins with two phases. The first is to use the final number of points on the convex envelope, which we denote by h, the exact value of which is clearly unknown at the beginning.

Phase 1:

- Partition the set S into $\frac{n}{h}$ subsets $\mathbb{B}_j, j = 1, .., \lceil \frac{n}{h} \rceil$, each containing at most h points (see Figure 3.7(a)).

- Use Graham's algorithm to calculate the convex hull for each subset $\mathbb{B}_j$ (see Figure 3.7(b)).

Phase 2:

- Eliminate points that are not selected in Phase 1 (see Figure 3.7(c)).

- Using Jarvis' algorithm, calculate the convex envelope of the remaining points (see Figure 3.7(d)).

In practice, since h is not known beforehand, the algorithm starts with a random value of h, which is often set to 5.

In [32], suggestions are made to improve the performance of the algorithm in practice:

- When calculating the convex envelopes of the subsets, it is possible to eliminate points that are not in those envelopes constituted in other iterations.

- When h increases, the new convex envelopes can be calculated by merging the previous ones to avoid their recalculation from zero.

The complexity of this algorithm is $O(n \log h)$, where h is the number of extreme points of the convex envelope [28].

3.2 Finding a concave hull of a set of points in the plane

There are many areas in which we are led to determine the shape of a set of points or of a graph. But in most cases, this form is not convex, hence the interest in determining such a shape in a more precise way using concave polygons.

Whereas the calculation of the convex envelope of a set of points is well explored by researchers, the calculation of a concave envelope is rarely studied. There are a few algorithms in the literature. The use of a graph's concave envelope is tantamount to looking for a cycle of the graph containing or circumscribing all of its vertices. The terms frontier or boundary are very commonly used to name such a cycle. Also observe, that such a cycle may pass

(a) 12 points in the plane

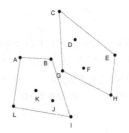

(b) Phase 1. Partition of the 12
disconnected points into 2
subsets, and application of
Graham's algorithm

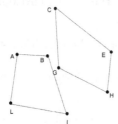

(c) Phase 2. Delete the
non-selected points in Phase 1

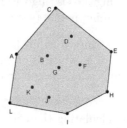

(d) The convex hull of the 12
points is formed by the set
$\{A, C, E, H, I, L\}$

Figure 3.7: An example for Chan's algorithm.

more than once through a vertex, which leads us to let $\mathbb{B}$ denote a sequence and not a set, as before.

To conclude this section and to be more complete on convex hull computation, we just mention the algorithm by Preparata and Hong [97] of complexity $O(n \log n)$, which is also based on the *divide and conquer* strategy and which extends to higher dimension, as well as the ultimate convex hull algorithm developed by Kirkpatrick and Seidel [72] in 1986 using the concept of *marriage before conquest*, which represents another output sensitive algorithm of complexity $O(n \log h)$.

3.2.1 Split and merge

Split and merge is an algorithm proposed in 1999 by Garai and Chaudhuri [49] to determine a concave envelope of a set of points. The name comes from the use of a splitting process followed by a fusion. In the split stage, some edges of the current convex envelope are replaced by others to include further points in the new boundary. The number of polygon edges resulting from this procedure can be set by the user. A major feature of this algorithm is that all polygon edges are independent of each other, which means in particular that it can be parallelized.

The complexity of this algorithm is $O(nm)$, where n is the number of the given points and m the number of edges of the resulting polygon.

The main phases of this algorithm can be described as follows:

Phase 1- Split procedure: in this phase, two points A and B, not yet explored, of the initial polygon are considered, and d is calculated, the average of the distances of the 3 closest neighbors of A and B which do not belong to the current polygon P. After that, two cases may arise:

- if $l \geq 1.5 * d$, then C, the nearest point to the line $[A, B]$, is added to the current polygon if there is no intersection between the lines $[A, C]$, $[B, C]$ and the edges of this polygon.

- otherwise, the algorithm chooses two other points and restarts the Split procedure from the beginning.

This phase is repeated until all edges of the polygon have been explored (see Figure 3.8).

Phase 2- Isolation procedure: for each ridge $\{x, y\}$ of the polygon found in the first step, the nearest 3 points are determined together with their average distance d to that edge.

If the distance $xy > d$, then select the edge $\{x', y'\}$ opposite to $\{x, y\}$ in P. If the quadrilateral $xyy'x'$ does not contain any point of S, then delete both $\{x, y\}$ and $\{x', y'\}$. This phase stops when all sides of the polygon have been visited, and it is only necessary if the set of points is separated into different subsets of isolated points.

Phase 3- Merge procedure: this phase begins with the result of Phase 1 if the points are not separated into dispersed subsets, and the result of Phase 2 otherwise. The first step is to calculate the internal angle of each vertex of the polygon. For each obtuse angle, $(\widehat{P_{i-1}P_iP_{i+1}} > 180)$, the area of the triangle $P_{i-1}P_iP_{i+1}$ is calculated. Now, select the minimum surface angle, remove the vertex P_i from the polygon and add the edge $\{P_{i-1}, P_{i+1}\}$.

The algorithm stops when the surface or the number of edges exceeds the user-defined values.

3.2.2 Perceptual boundary extraction

This approach has been proposed by Chaudhuri et al. [33] in 1997 for the search of a concave hull of a set of points. It is based on a concept called s-shape which creates an edge resembling a staircase, and then uses another parameter calculated on the basis of the s-shape. Its complexity is $O(n)$.

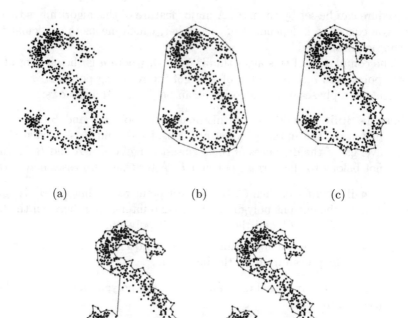

(a)   (b)   (c)

(d)   (e)

Figure 3.8: The progress of the Split procedure from convex to concave hull.

The algorithm can be summarized in three steps:

1. Select the parameter s and calculate the corresponding parameter r.

2. Generate a corresponding r-shape.

3. Clean the inconsistent parts of the form generated by the previous step.

Like Split and Merge, Perceptual boundary extraction can be implemented on parallel machines since the operations are local at each point.

3.2.3   K-nearest neighbor

This algorithm was developed by Moreira and Santos [90] in 2007 for the determination of an envelope (convex or non-convex) of a set of points in the plane. Their approach is based on the concept of the k nearest neighbors of a point, where k is a user-controlled parameter. The authors have proposed this algorithm to calculate the boundary of a geographic area defined by a set of points in a given region. These points represent the POIs (Points of Interest) that are an integral part of geographic databases and navigation systems.

The k-nearest neighbor (or KNN) algorithm aims to select the points that form a polygon delimiting an arbitrary set of points in a plane. We note that this procedure does not generate a unique solution (polygon), but depends on the input parameter k that fixes the number of considered neighbors.

The approach is based on Jarvis' algorithm, except that instead of considering each time the minimum polar angle with respect to all points of the set, the minimum angle is calculated only with the k nearest points.

First of all, we find a point P_0 with minimum y-coordinate (located at the bottom of the point set). To determine the next vertex that will be part of the polygon, we consider the k nearest points $P_j, j = 1, 2, ..., k$ and we measure the angle formed by each edge $\{P_0, P_1\}, \{P_0, P_2\}, ..., \{P_0, P_k\}$ with the horizontal axis. The point with the minimum angle is selected, and the line segment linking the current point to this newly selected point is added. The same process is repeated with the new point taken as the current one until the first P_0 point is selected again.

We notice that the greater the number k, the smoother the resulting polygon. If $k = n$, then KNN behaves exactly like Jarvis' algorithm, which means that we obtain the convex envelope.

Two cases, where this algorithm does not work, have been reported by the authors of [90]:

1. an added ridge crossing an existing ridge is likely to find points that are not boundary points.

2. all the points are distributed over several isolated regions. In this case, we only find the convex hull of the points situated in the area containing P_0.

It should be noted that the two cases mentioned above can be avoided by increasing the parameter k appropriately.

3.2.4 Concaveness measure

Park and Oh [93] propose an algorithm to calculate the concave envelope, which can also be applied in dimension $p > 2$. The authors propose a concavity measure and a graph that captures the geometric shape of a data set.

The principle of this algorithm consists in using a well-known algorithm to calculate the convex envelope and then to *"dig"* the result to determine a concave envelope with appropriate depth. The only input parameter for this algorithm is a threshold ε which corresponds to the excavation level. The smaller ε, the more accurate the result. For a certain threshold ε big enough, we get the convex envelope, because the algorithm will not execute the second *"dig"*-step.

The algorithm consists of the following three major steps:

1. Determine the convex envelope and the threshold ε.

2. Calculate the nearest points around the boundary, as well as the minimum distance L to be used as the "decision distance".

3. If $\frac{\text{length of edge}}{\text{decision distance}} > \varepsilon$ then the excavation process is carried out and repeated from the second step onwards.

3.3   Finding a polygon hull of a Euclidean graph

Finding the boundary of a graph is not an easy task, and together with the connectivity constraint an additional difficulty arises from the case discussed above that only edges of the graph can be used to connect consecutive vertices. A main objective of this book is to present algorithms for the three problems introduced in the previous chapter, i.e., A-polygon hull, B-polygon hull and C-polygon hull. The distributed versions of these algorithms will be presented in a subsequent chapter.

3.3.1   LPCN: Least Polar-angle Connected Node algorithm to find a polygon hull of a connected Euclidean graph

The search for a polygonal envelope of a connected Euclidean graph can be assimilated to that of a convex envelope of a set of points, with the restriction that only the edges of the graph can be used for this envelope. This last one has the form of a pseudo-polygon (cf. Section 1.10.2).

The algorithm that we will present in this section is inspired by Jarvis in the sense that at each iteration we look for the vertex that forms the minimum polar angle among the neighbors of the current vertex. Our algorithm has a complexity of $O(dh)$, where d is the maximum degree of the graph and h the number of vertices on the polygonal envelope.

Let us consider a connected Euclidean graph $G = (V, E)$, where V is the set of n vertices and E the set of m edges. The first version of the LPCN algorithm (Least Polar-angle Connected Node) proposed by Bounceur et al. in [28, 76] is able to find the polygonal envelope for a connected Euclidean graph, with the exception of some critical cases that are analysed and resolved by a refined version that has also appeared in [76].

The LPCN algorithm can be described as follows, where, based on Section 1.11.3.2, the algorithm stops when the starting vertex visits a second time the first visited neighbor. This means that the starting angle is visited twice. It is also possible to use another stop condition based on the sum of the visited angles of the polygon hull until reaching the starting vertex a second time. As shown in Section 1.11.3.1, this sum must be equal to $360°$.

Algorithm 5: LPCN1

Data: V, E

Result: $\mathbb{B}, \mathbb{B}_E$

1 $P_c \leftarrow$ a vertex with minimum x-coordinate;

2 $P_{first} \leftarrow P_c$;

3 $P_p \leftarrow$ a fictitious vertex situated to the left of P_c;

4 $\mathbb{B} \leftarrow \{P_c\}$;

5 $\mathbb{B}_E \leftarrow \emptyset$;

6 **repeat**

7 $\quad P_v \leftarrow \underset{P_j \in N(P_c)}{\operatorname{argmin}} \{\varphi(P_p, P_c, P_j)\}$;

8 $\quad \mathbb{B} \leftarrow (\mathbb{B}, P_v)$; $\mathbb{B}_E \leftarrow \mathbb{B}_E \cup \{\{P_c, P_v\}\}$;

9 $\quad P_p \leftarrow P_c$;

10 $\quad P_c \leftarrow P_v$;

11 **until** $(P_p = P_{first}$ *and* $P_v \in \mathbb{B})$;

3.3.2   Concave hull of a plane Euclidean graph

Now let us consider the plane Euclidean graph as shown in Figure 3.9 and explain step-by-step how Algorithm 5 works on this example. The algorithm starts with the minimum x-coordinate vertex $P_c = v_6$. Then it adds a fictitious vertex P_p situated to the left of P_c, and calculates the minimum angle among the angles formed by the neighbors of P_c which are: $\widehat{P_p v_6 v_7}, \widehat{P_p v_6 v_8}, \widehat{P_p v_6 v_5}$. In this case, the chosen angle is $\widehat{P_p v_6 v_7}$. The same procedure is applied to $P_c = v_7$. We notice that in the presence of a vertex with degree 1, in our example v_{13}, we have to include its neighbor v_{12} a second time, as illustrated in Figure 3.9(h).

3.3.3   Concave hull of a general Euclidean graph

In many real-world applications, a representation of a graph contains edge-crossings or intersections. This situation may pose a problem for Algorithm 5 presented above, in particular, if the intersection occurs at the boundary of the graph. In this section, we present algorithms that solve the A-polygon, B-polygon and C-polygon hull problems that we already formulated in Section 2.4.3.

3.3.3.1 A-polygon hull

This problem consists in looking for a set of vertices forming the boundary of the graph, i.e., a cycle of the graph such that all vertices of the graph are either on or circumscribed by this cycle, in mathematical terms:

$$\mathbb{V}_a \subseteq V \text{ and } v_i \in \widehat{P}(\mathbb{V}_a), \forall v_i \in V \tag{3.1}$$

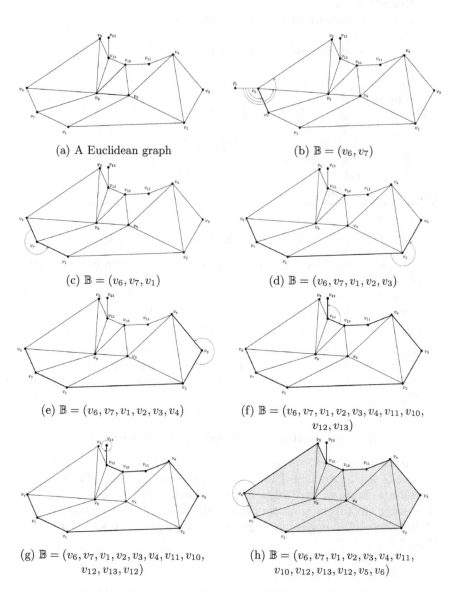

(a) A Euclidean graph

(b) $\mathbb{B} = (v_6, v_7)$

(c) $\mathbb{B} = (v_6, v_7, v_1)$

(d) $\mathbb{B} = (v_6, v_7, v_1, v_2, v_3)$

(e) $\mathbb{B} = (v_6, v_7, v_1, v_2, v_3, v_4)$

(f) $\mathbb{B} = (v_6, v_7, v_1, v_2, v_3, v_4, v_{11}, v_{10},$ $v_{12}, v_{13})$

(g) $\mathbb{B} = (v_6, v_7, v_1, v_2, v_3, v_4, v_{11}, v_{10},$ $v_{12}, v_{13}, v_{12})$

(h) $\mathbb{B} = (v_6, v_7, v_1, v_2, v_3, v_4, v_{11},$ $v_{10}, v_{12}, v_{13}, v_{12}, v_5, v_6)$

Figure 3.9: An example for the LPCN1 algorithm.

Algorithm 6: LPCN2

Data: V, E

Result: $\mathbb{B}_V$, $\mathbb{B}_E$

1 $P_0 \leftarrow P_c \leftarrow$ a vertex with minimum x-coordinate;

2 $P_p \leftarrow$ a fictitious vertex situated to the left of P_c;

3 $\mathbb{B}_V \leftarrow (P_c)$;

4 $\mathbb{B}_E \leftarrow \emptyset$;

5 *once* $\leftarrow true$;

6 **repeat**

7      $\mathbb{A} \leftarrow \{P \in N(P_c) / \mathbb{B}_E \cap \{\{P_c, P\}\} = \emptyset\}$;

8      $P_{min} \leftarrow \underset{P \in \mathbb{A}}{\operatorname{argmin}}\{\varphi(P_p, P_c, P)\}$;

9      $\mathbb{B}_V \leftarrow (\mathbb{B}_V, P_{min})$;

10      $\mathbb{B}_E \leftarrow \mathbb{B}_E \cup \{\{P_c, P_{min}\}\}$;

11      $P_p \leftarrow P_c$;

12      $P_c \leftarrow P_{min}$;

13   **if** $(once = true)$ **then**

14   *once* $\leftarrow false$;

15          $P_{first} \leftarrow P_{min}$

16   **end**

17 **until** $((P_c = P_0) \; and \; (P_{min} = P_{first}))$;

We can see in Figure 3.10 how this second LPCN-version works.

3.3.3.2 B-polygon hull

The number of vertices found by Algorithm 6 may need to be minimized in order to minimize the cost of using this boundary. The mathematical formulation of this problem is given by the Model 2.11 (see Section 2.4.3).

We propose a new approach to solve this problem. Our procedure is based on the LPCN algorithm and is described in Algorithm 7, where:

$\mathbb{H}_V \subseteq V$: vertex sequence describing the path linking $\mathbb{P}_h$ to $\mathbb{P}_v$.

$\mathbb{H}_E \subseteq E$: set of edges describing the path linking $\mathbb{P}_h$ to $\mathbb{P}_v$.

In this algorithm, we use the polar angle principle used by LPCN. We need to define temporary sets $(\mathbb{H}_V, \mathbb{H}_E)$, which are used to count the two paths in both directions of the polar angle and then choose the one that has minimal cardinality. At each time a crossing situation occurs between a border edge and a candidate edge we use the sets $\mathbb{H}_V$ and $\mathbb{H}_E$ to store the chain obtained by executing the algorithm in the opposite direction of the polar angle until another crossing situation appears.

We can see in Figure 3.11 how LPCN3 works on an example.

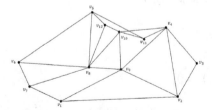

(a) A connected Euclidean graph

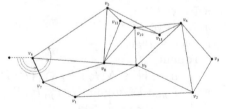

(b) $\mathbb{B}_V = (v_6, v_7)$

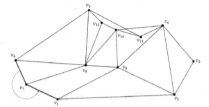

(c) $\mathbb{B}_V = (v_6, v_7, v_1)$

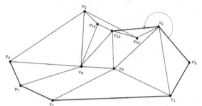

(d) $\mathbb{B}_V = (v_6, v_7, v_1, v_2, v_3, v_4, v_{10})$

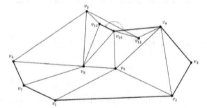

(e) $\mathbb{B}_V = (v_6, v_7, v_1, v_2, v_3, v_4, v_{10}, v_{12})$

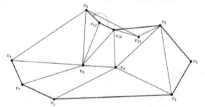

(f) $\mathbb{B}_V = (v_6, v_7, v_1, v_2, v_3, v_4, v_{10}, v_{12}, v_5)$

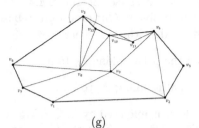

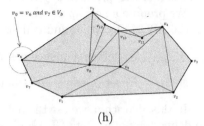

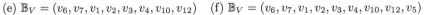

(g)
$\mathbb{B}_V = (v_6, v_7, v_1, v_2, v_3, v_4, v_{10}, v_{12}, v_5, v_6)$

(h)
$\mathbb{B}_V = (v_6, v_7, v_1, v_2, v_3, v_4, v_{10}, v_{12}, v_5, v_6)$

Figure 3.10: An example for LPCN2.

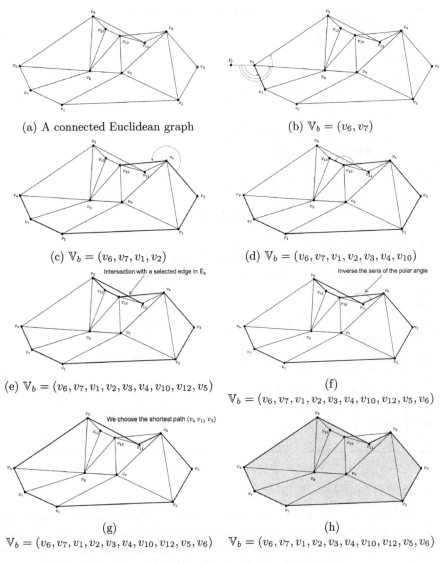

(a) A connected Euclidean graph

(b) $\mathbb{V}_b = (v_6, v_7)$

(c) $\mathbb{V}_b = (v_6, v_7, v_1, v_2)$

(d) $\mathbb{V}_b = (v_6, v_7, v_1, v_2, v_3, v_4, v_{10})$

(e) $\mathbb{V}_b = (v_6, v_7, v_1, v_2, v_3, v_4, v_{10}, v_{12}, v_5)$

(f)
$\mathbb{V}_b = (v_6, v_7, v_1, v_2, v_3, v_4, v_{10}, v_{12}, v_5, v_6)$

(g)
$\mathbb{V}_b = (v_6, v_7, v_1, v_2, v_3, v_4, v_{10}, v_{12}, v_5, v_6)$

(h)
$\mathbb{V}_b = (v_6, v_7, v_1, v_2, v_3, v_4, v_{10}, v_{12}, v_5, v_6)$

Figure 3.11: An example for LPCN3.

Algorithm 7: LPCN3

Data: V, E

Result: $\mathbb{V}_b$, $\mathbb{E}_b$

1 $P_0 \leftarrow P_c \leftarrow$ a vertex with minimum x-coordinate;

2 $P_p \leftarrow$ a fictitious vertex situated to the left of P_c;

3 $\mathbb{V}_b \leftarrow \{P_c\}$;

4 $\mathbb{E}_b \leftarrow \emptyset$;

5 *once $\leftarrow$ true*;

6 **repeat**

7    $P_v \leftarrow \underset{P \in N(P_c)}{\operatorname{argmin}} \{\varphi(P_p, P_c, P)\}$;

8   **if** $\mathbb{E}_b \cap \{(P_c, P_v)\} \neq \emptyset$ **then**

9      $P_h \leftarrow P_c$; $\mathbb{H}_V = \{P_c\}$; $\mathbb{H}_E = \emptyset$;

10   **while** $P_v \notin \mathbb{V}_b$ **do**

11        $\mathbb{H}_V \leftarrow (\mathbb{H}_V, P_v)$;

12        $\mathbb{H}_E \leftarrow \mathbb{H}_E \cup \{(P_c, P_v)\}$;

13        $P_p \leftarrow P_c$; $P_c \leftarrow P_v$;

14        $P_v \leftarrow \underset{P \in N(P_c)}{\operatorname{argmax}} \{\varphi(P_p, P_c, P)\}$;

15   **end**

16   Let $\mathbb{C}_V, \mathbb{C}_E$ be a chain from $\mathbb{V}_b$ and $\mathbb{E}_b$ linking P_v to P_h;

17   **if** $|\mathbb{V}_b| < |\mathbb{H}_V|$ **then**

18        $\mathbb{V}_b = \mathbb{V}_b \backslash \mathbb{C}_V$;

19        $\mathbb{V}_b = (\mathbb{V}_b, \mathbb{H}_V)$;

20        $\mathbb{E}_b = \mathbb{E}_b \backslash \mathbb{C}_E \cup \mathbb{H}_E$;

21   **end**

22   **else**

23      $\mathbb{V}_b \leftarrow (\mathbb{V}_b, P_v)$;

24      $\mathbb{E}_b \leftarrow \mathbb{E}_b \cup \{(P_c, P_v)\}$;

25      $P_p \leftarrow P_c$; $P_c \leftarrow P_v$;

26   **end**

27 **until** $((P_c = P_0)$ *and* $(P_v \in \mathbb{V}_b))$;

3.3.3.3   *C*-polygon hull

As already mentioned in the previous chapter, this problem is twofold. An ideal solution for graphs with intersections on the boundary does not exist. That is why we will use compromise solutions, and formulate the problem as a bi-objective optimization problem (see Problem 2.12). To solve this Problem, we propose an interactive method using the LPCN algorithm together with a weighting of the two criteria by real weights λ_1 and λ_2, these weights being part of the input parameters.

We note that Algorithm 8 is used to find a compromise solution to Problem (2.12). It is mainly based on LPCN1, with a modification in the choice of

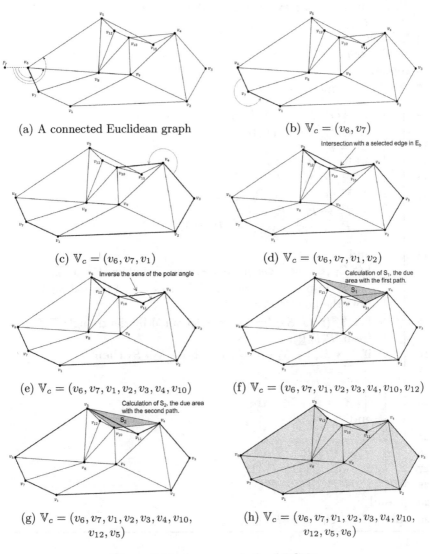

(a) A connected Euclidean graph

(b) $\mathbb{V}_c = (v_6, v_7)$

(c) $\mathbb{V}_c = (v_6, v_7, v_1)$

(d) $\mathbb{V}_c = (v_6, v_7, v_1, v_2)$

(e) $\mathbb{V}_c = (v_6, v_7, v_1, v_2, v_3, v_4, v_{10})$

(f) $\mathbb{V}_c = (v_6, v_7, v_1, v_2, v_3, v_4, v_{10}, v_{12})$

(g) $\mathbb{V}_c = (v_6, v_7, v_1, v_2, v_3, v_4, v_{10}, v_{12}, v_5)$

(h) $\mathbb{V}_c = (v_6, v_7, v_1, v_2, v_3, v_4, v_{10}, v_{12}, v_5, v_6)$

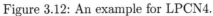

Figure 3.12: An example for LPCN4.

Algorithm 8: LPCN4

Data: V, E, Λ

Result: $\mathbb{V}_c, \mathbb{E}_c$

1 $P_0 \leftarrow P_c \leftarrow$ a vertex with minimum x-coordinate;

2 $P_p \leftarrow$ a fictitious vertex situated to the left of P_c;

3 $\mathbb{V}_c \leftarrow \{P_c\}$;

4 $\mathbb{E}_c \leftarrow \emptyset$;

5 *once* $\leftarrow$ *true*;

6 **repeat**

7 $\quad$ $P_v \leftarrow \underset{P \in N(P_c)}{\operatorname{argmin}} \{\varphi(P_p, P_c, P)\}$;

8 $\quad$ **if** $\mathbb{E}_c \cap \{\{P_c, P_v\}\} \neq \emptyset$ **then**

9 $\quad\quad$ $P_h \leftarrow P_c$; $\mathbb{H}_V = \{P_c\}$;

10 $\quad\quad$ $\mathbb{H}_E = \emptyset$;

11 $\quad\quad$ **while** $P_v \notin \mathbb{V}_c$ **do**

12 $\quad\quad\quad$ $\mathbb{H}_V \leftarrow (\mathbb{H}_V, P_v)$;

13 $\quad\quad\quad$ $\mathbb{H}_E \leftarrow \mathbb{H}_E \cup \{\{P_c, P_v\}\}$;

14 $\quad\quad\quad$ $P_p \leftarrow P_c$; $\quad P_c \leftarrow P_v$;

15 $\quad\quad\quad$ $P_v \leftarrow \underset{P \in N(P_c)}{\operatorname{argmax}} \{\varphi(P_p, P_c, P)\}$;

16 $\quad\quad$ **end**

17 $\quad\quad$ Let $\mathbb{C}_V, \mathbb{C}_E$ be a chain from $\mathbb{V}_c$ and $\mathbb{E}_c$ linking P_v to P_h;

18 $\quad\quad$ Let $\overline{P}$ (resp. $\underline{P}$) be a vertex of $\mathbb{H}_V$ (resp. $\mathbb{C}_V$) located farthest from the line $[P_v P_h]$;

19 $\quad\quad$ Let S_1 (resp. S_2) be the area delimited by the triangle $P_v P_h \overline{P}$ (resp. $P_v P_h \underline{P}$);

20 $\quad\quad$ **if** $\lambda_1 * |\mathbb{C}_V| + \lambda_2 * S_1 > \lambda_1 * |\mathbb{H}_V| + \lambda_2 * S_2$ **then**

21 $\quad\quad\quad$ $\mathbb{V}_c = \mathbb{V}_c \backslash \mathbb{C}_V$;

22 $\quad\quad\quad$ $\mathbb{V}_c = (\mathbb{V}_c, \mathbb{H}_V)$;

23 $\quad\quad\quad$ $\mathbb{E}_c = \mathbb{E}_c \backslash \mathbb{C}_E \cup \mathbb{H}_E$;

24 $\quad\quad$ **end**

25 $\quad$ **else**

26 $\quad\quad$ $\mathbb{V}_c \leftarrow (\mathbb{V}_c, P_v)$;

27 $\quad\quad$ $\mathbb{E}_c \leftarrow \mathbb{E}_c \cup \{\{P_c, P_v\}\}$;

28 $\quad\quad$ $P_p \leftarrow P_c$; $\quad P_c \leftarrow P_v$;

29 $\quad$ **end**

30 **until** $((P_c = P_0)$ *and* $(P_v \in \mathbb{V}_c))$;

---

a path at a boundary intersection and with two objectives: maximize the area occupied by the graph, and minimize the number of vertices on the boundary. The first objective is similar to the search for a path that reduces the area by a part of the graph that will not be counted if we follow the path, which explains the weighting of the two objectives each time such a situation arises.

In the example of Figure 3.12, we have chosen $\lambda_1 = \lambda_2 = 0.5$ for objective functions f_1 and f_2. Thus, in Figure 3.12(f) and Figure 3.12(g), we have the choice to select either the path $\mathbb{H}_V = (v_4, v_{11}, v_5)$ or $\mathbb{C}_V = (v_5, v_{12}, v_{10}, v_{11}, v_4)$. At first, we find the vertices that are at maximum distance from the line $[v_4, v_5]$. These two vertices are: $v_{11} \in \mathbb{H}_V$ and $v_{10} \in \mathbb{C}_V$. Then we calculate the area lost by each of the paths: S_1 (respectively S_2) which corresponds to the triangle $v_4 v_{11} v_5$ (resp. $v_4 v_{10} v_5$). Finally, two quantities are calculated: $Q_1 = \lambda_2|\mathbb{H}_V| + \lambda_2 S_1$ and $Q_2 = \lambda_1|\mathbb{C}_V| + \lambda_2 S_2$; as the quantity Q_2 is smaller than Q_1, we choose the path $\mathbb{C}_V$.

3.4 Finding the polygon hull of a Euclidean graph without conditions on the starting vertex

All the versions of the LPCN algorithm presented in the previous section start from a known vertex, which is the one having the smallest x-coordinate. In real applications, it is possible to find situations where the graph is constructed dynamically and where the global minimum cannot be determined at the beginning. For example, drawing the contours of a zone of interest in an image. In this case, the medical practitioner will choose manually any pixel inside of the zone of interest from which it is possible to determine the first form of the graph based only on its neighbors. The complete graph is constructed after many iterations where in each iteration the next neighbors are determined based on the previously found neighbors. LPCN must be adapted to this situation. To do this, we propose a new concept called *Reset and Restart* which will be combined with the LPCN algorithm. This will lead to an algorithm that allows to run the LPCN from any vertex of the graph. We may call it *Reset and Restart Least Polar-angle Connected Node (RRLPCN)* [24].

This RRLPCN algorithm starts the LPCN algorithm from a given vertex of the graph which is not necessarily a boundary vertex. This vertex will be assumed as the one having the minimum x-coordinate x_{min}. Then in each iteration another boundary vertex is determined using the angles, as explained in the previous section, until all boundary vertices have been visited. If in a given iteration k the determined boundary vertex v_k has an x-coordinate smaller than x_{min}, then we will stop visiting the other vertices and all the already determined boundary vertices will be considered as non-boundary vertices *(Reset)* and the LPCN algorithm will be restarted from the vertex v_k *(Restart)*.

To illustrate the Reset and Restart concept, for simplicity, let us take the graph shown in Figure 3.13, where the vertex with the minimum x-coordinate is linked by a chain to the starting vertex. The marking here is of no particular sense, it is just to highlight the visited vertices. However, it has a sense in the case where the visited vertices have to be marked as in the case of determining the boundary vertices as explained below (see Figure 3.13).

Suppose that the algorithm starts from vertex 1 as shown by Figure 3.13(a). First, we will mark it and set the value of $x_{min} = x_1$, the x-coordinate of vertex 1 and go to vertex 2 as shown by Figure 3.13(b) and mark it. Since the x-coordinate of vertex 2 is greater than x_{min}, we will not do anything. Then, we will go to vertex 3, mark it, and compare its x-coordinate to x_{min} (see Figure 3.13(c)). Again, the x-coordinate of vertex 3 is also greater than x_{min}. Now, we will go to vertex 4 (see Figure 3.13(d)) and mark it. Here, however, the x-coordinate of vertex 4 is less than x_{min}. In this case, we execute the *Reset and Restart* process. First, we will *Reset* by unmarking the vertices 1, 2 and 3 (Figure 3.13(e)), *Restart* the algorithm from vertex 4 (Figure 3.13(f)), and update the value of $x_{min} = x_4$, the x-coordinate of vertex 4. Then, we will go to vertex 5 as shown by Figure 3.13(g) and mark it. Here also, since the x-coordinate of vertex 5 is less than x_{min}, we will execute the *Reset and Restart* process. First, we will *Reset* by unmarking vertex 4 (Figure 3.13(h)) and we stop the algorithm since there is no other vertex to visit (Figure 3.13(i)). Finally, the marked vertex 5 is the one having the smallest x-coordinate in the chain.

The following steps show how to combine the *Reset and Restart* concept with the LPCN algorithm:

1) In the *Restart* step, the starting vertex will select the next vertex. The next vertex is the one which forms the minimum polar angle between the starting vertex, a fictitious vertex situated to its left and its neighbors, as shown by Figure 3.14. In this figure, the edge between the starting vertex and the next vertex is colored gray.

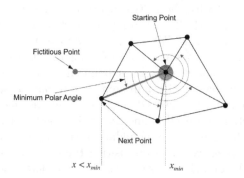

Figure 3.14: Calculating the minimum polar angle from any starting vertex.

2) Once the next vertex is selected, we have two possible cases: Either the x-coordinate of this vertex is greater or equal to x_{min} or it is smaller than x_{min}. In the first case, we continue the execution of LPCN and

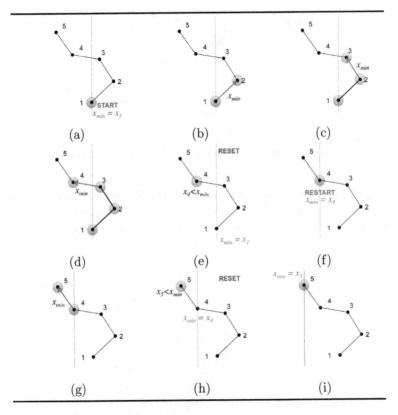

Figure 3.13: *Reset and Restart* concept.

select the next vertex. In the second case, we execute the *Reset* and set x_{min} equal to the actual x-coordinate and *Restart* the algorithm from this vertex. Figure 3.15 shows an example in which all the x-coordinates of the selected vertices are smaller than x_{min}.

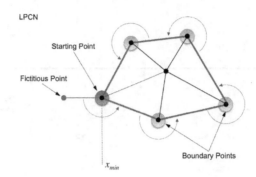

Figure 3.15: An example for finding a polygon hull.

Algorithm 9 shows the pseudo-code of the Reset and Restart Least Polar-angle Connected Node (RRLPCN) that allows to find the boundary vertices by starting from any vertex of a connected Euclidean graph.

Algorithm 9: RRLPCN

Data: V, E

Result: $\mathbb{B}_V, \mathbb{B}_E$

1 $P_c \leftarrow$ an arbitrary vertex chosen from V;
2 $\mathbb{B}_V \leftarrow [P_c]$;
3 $P_{first} \leftarrow P_c$;
4 $P_p \leftarrow$ a fictitious vertex situated to the left of P_{first};
5 $X_{first} \leftarrow$ the x-coordinate of P_{first};
6 **repeat**
7      $\mathbb{A} \leftarrow \emptyset$;
8      $P_v \leftarrow \underset{P_j \in N(P_c) \ \& \ P_j \notin \mathbb{A}}{\operatorname{argmin}} \{\varphi(P_p, P_c, P_j)\}$;
9      $X_v \leftarrow$ the x-coordinate of P_v;
10   **if** $X_v < X_{first}$ **then**
11   *Reset:* $\mathbb{B}_V \leftarrow \emptyset$; $\mathbb{B}_E \leftarrow \emptyset$;
12   *Restart:* $P_c \leftarrow P_v$; go to step 3;
13   **end**
14   **if** $\mathbb{B}_E \cap \{P_c, P_v\} \neq \emptyset$ **then**
15          $\mathbb{A} \leftarrow \mathbb{A} \cup \{P_v\}$;
16   Go to 9;
17   **end**
18      $\mathbb{B}_V \leftarrow (\mathbb{B}_V, P_v)$;
19      $\mathbb{B}_E \leftarrow \mathbb{B}_E \cup \{\{P_c, P_v\}\}$;
20      $P_p \leftarrow P_c$;
21      $P_c \leftarrow P_v$;
22 **until** $P_v = P_{first}$;

Chapter 4

Distributed algorithms for boundary detection

Sometimes the notions of *sequential*, *parallel* and *distributed* algorithms are confusing. Even if the difference between sequential and parallel algorithms can be understood easily, distinguishing parallel from distributed algorithms can remain ambiguous, especially when we only start to know these areas. In the first part of this chapter, we try to clarify the difference between these concepts by answering the question: *"What is a distributed algorithm?"*

The second part of this chapter is dedicated to the presentation of some basic distributed algorithms with special emphasis on those that are required for finding polygon hulls. Then we will present several new algorithms allowing to elect in a network of distributed nodes or systems a particular node with respect to an associated value. This topic, also called *leader election*, is essential for distributed systems since their architecture is not centralized, and it is not easy to find, for instance, the node having the minimum value.

On the basis of these concepts, we will conclude this chapter with a presentation of the distributed versions of the LPCN and the RR-LPCN algorithms introduced in the previous chapter.

4.1 What is a distributed algorithm?

Before presenting the concept of distributed algorithms, let us discuss the concept of centralized (sequential) and parallel algorithms, and indicate how they differ from distributed algorithms. In a centralized program, the instructions are executed one after the other, i.e., sequentially. In a parallel program, the instructions are executed in parallel. To accelerate the execution of a program, the interest lies in finding those parts of it that can be parallelized.

As an example, let us consider the following instructions on variables x (supposed to be initialized at value 0) and y:

```
1: x = x + 1
2: y = 2 * x
3: print x
```

```
4: print y
```

The sequential execution of this program will display 1 for x and 2 for y, whereas its parallel execution will display 1 for x and 0 for y. However, it is possible to parallelize lines 3 and 4 only, in which case we obtain the same result as in the sequential execution.

The parallelization can be performed by executing each part of the program on different processes of an operating system. It can be performed using GPU (*Graphical Processing Unit*) where the parallelization is already designed materially and the program must be adapted to its architecture. We call this kind of parallelization *massively parallel*.

Another way to do parallelization is to integrate it in the programming language, a concept called *multi-threading*. Here only the threads are executed in parallel, their instructions, however, in a sequential way. If in the previous example, the instructions 1 and 2 are placed inside a thread, as shown by the following example, then they will be executed sequentially but in parallel with the other instructions (5 and 6). The sequential execution of the following program will display 0 for x and 0 for y but at the end of the program the value of x will be 1, and the one of y will be 2.

```
1: thread {
2:    x = x + 1
3:    y = 2 * x
4: }
5: print x
6: print y
```

The main issue of parallel and multi-threading programming is the synchronization. Let us consider the following program:

```
1: thread {
2:    x = x + 1
3:    y = 2 * x
4: }
5: thread {
6:    delay(1)
7:    print x
8:    print y
9: }
```

Lines 2 and 3 are executed at the same time as lines 6, 7 and 8. In this case, the second thread will wait for 1 millisecond, while the first thread is calculating the values of x and y. Then it displays the values 1 and 2 as in the sequential program. However, it is not sure that the execution will be done as this in terms of execution time. There is no guarantee that the calculation of x and y will terminate during the waiting time of 1 millisecond. To obtain such a guarantee, it is necessary to add some testing. For example, we can add a boolean variable b and replace the waiting time by a *while* loop testing whether b is true or not, and in the first thread we modify the value of b once x and y are calculated. This operation is called *synchronization*. To do this in an optimal way, any programming language integrates a mechanism and some instructions to manage synchronization, because using just the while instruction is not reliable.

The concept of distributed algorithms cannot be used for programs. This means that a program cannot be distributed. However, one can use the term distributed program for a program that is executed in a distributed system, machine, etc. In such a case, each system will execute its own program independently from the others. The program of each system can then be executed sequentially, in parallel, in a thread, etc. Any distributed system can communicate or interact with each other through messages. This communication gives the possibility to design and implement some algorithms in a distributed way in order to achieve a common goal. For instance, we can implement the distributed shortest path algorithm in order to route an information from a system to another. Two systems are not necessarily communicating. This can be due to the distance between them or to a lack of connection between them.

Distributed systems can be synchronous or asynchronous. In the synchronous case, each system will execute a set of instructions (actions) during a given time depending on the local clock. This time, called a round, is the same for each system. Once a system has executed an action, it goes to the next action which will be executed in the next round, and so on. It is also assumed that any sent message will be received before the end of the current round. In the asynchronous case, the systems are totally independent. As in parallel programming, it is possible to introduce some synchronization by adding special systems (synchronizers) or by implementing some codes within existing systems. However, adding synchronization to asynchronous distributed systems will not make them synchronous.

Systems that belong to the same machine, for example, cores, NoCs (Networks on Chip), cannot be considered as distributed systems since they have access to a shared memory and they share the same clock.

Throughout this chapter, we restrict ourselves to asynchronous distributed systems in which the instructions of each system are executed sequentially. The following programs provide examples of two independent programs injected into two distributed systems, where the first one allows the first system to send each second a random value to the second system, and the second system calculates the sum of the received values and displays it.

```
//System 1
1: x = random()
2: send(x, 2) // Send x to System 2
```

```
//System 2
1: v = read()
2: s = s + v
3: print s
```

Finally, distributed systems are characterized with respect to parallel machines by:

- a great number of computation entities (systems),

- physical distance between the systems.

We may consider distributed calculation as the art of doing calculations and running algorithms on distributed systems, the main objectives being:

- to increase the speed of calculation and data storage,

- to allow communication and share data (distributed data) among several distant systems.

Finally, in the case of Big Data, where a huge number of data is continually received and analyzed on the internet, data scientists use distributed servers situated anywhere in the world, so-called data centers, to execute their calculations and algorithms in a parallel way.

4.2   Basic concepts

In this section, we will introduce some basic concepts and notions that are to be used throughout this chapter. Since we are dealing with the main problem of finding hulls of graphs, and especially polygon hulls, we will use the term *network* to denote a Euclidean graph and the term *node* to designate a vertex of a network. Networks are widely used to represent distributed systems, like computers, smartphones, unmanned areal vehicles (UAVs), Arduino cards, Raspberry cards, etc. Figure 4.1 shows an example of a random network.

A direct link between two nodes indicates the possibility of two systems to directly communicate with each other, communication that we assume to be symmetrical. In this case, a sensor network can be modeled as an undirected graph $G = (S, E)$, where $S = \{s_1, s_2, ..., s_n\}$ represents the sensor nodes, $n = | S |$ their total number and E the set of communication links. A link between two nodes can be defined as follows:

$$e_{ij} = \begin{cases} 1 & \text{if node } s_i \text{ directly communicates with node } s_j, \\ 0 & \text{otherwise.} \end{cases} \tag{4.1}$$

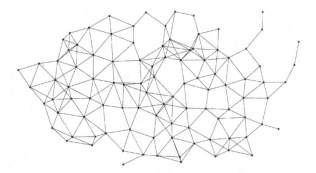

Figure 4.1: A random network.

The neighborhood $N(s_i)$ of a node s_i is the set of nodes that directly communicate with it.

$$N(s_i) = \{s_j / e_{ij} = 1, j = 1, .., n \text{ and } j \neq i\} \tag{4.2}$$

A node can communicate by sending and receiving messages. As shown by Figure 4.2, it can use a unicast (direct) or a broadcast transmission.

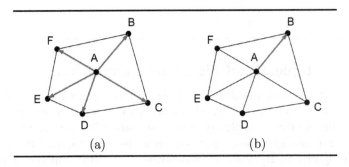

Figure 4.2: Message transmission in a network.

In the unicast transmission (Figure 4.2(a)), the transmitter uses the instruction `send(m,id)` to send a direct message m to the system having the identifier `id` and the instruction `read()` to read received messages. This last instruction is blocking, that is to say, if no messages are received any more, the instructions following it will not be executed. However, it is possible to add a limited time to continue the execution after a given blocking time. In this case, the instruction must be written as `read(t)`, and if no message is received during a time period t, the system will continue the execution of the following instructions of the program.

In the broadcast transmission (Figure 4.2(b)), a system can send only one message to its neighbors using the instruction `send(m,*)`. The neighbors of a transmitter represent all systems that can communicate with it. In terms of radio, the neighbors are all those systems that are within the radio range of the transmitter. In real-world systems, even a receiver is inside the radio range of a transmitter, it is not guaranteed to receive its messages. This is mainly due to propagation and interferences. Propagation represents the power of the received signal at the place where the receiver is located. Interference can affect the received signal because of the presence of systems around. If their number is important or if they are communicating at the same time, it is possible to affect the communication and to cause a loss of messages. To deal with this problem, the systems can use acknowledgment (ACK) messages where for each sent message the transmitter waits a given time for an acknowledgment message. If no acknowledgment is received, the operation will be reiterated, and if after three attempts no acknowledgment has been received, the transmitted message is considered as lost.

Finally, distributed systems are, as any electronic device, subject to failures. The main problem in this case is that if a system fails then all the connected systems will be obsolete and the network can be considered as nonactive or dead. That is why distributed algorithms must deal with failures and have to be robust and fault tolerant. In this case, even if a system fails, the network continues to work and to execute the algorithm in a normal way.

4.3   Complexity of distributed algorithms

In order to compare distributed algorithms, we need to evaluate their performance. Several metrics have been proposed for such an evaluation [83, 110], including the time complexity whose estimation is relatively easy in the case of synchronous systems or networks. It is the time required for the execution of an algorithm. However, in asynchronous systems, different types of communication between systems exist, which makes the estimation of the communication complexity more intricate, reason for which it is considered as the main performance measure in this context [67, 85].

Sometimes, distributed systems are designed to centralize the information in one system called a sink. In this case, the calculation is done by that single system. It is then possible to use the classical notion of complexity which is based on the number of elementary operations.

In the case of communication complexity, we estimate the *number of exchanged messages*. Since the algorithms presented in this book are especially designed for Wireless Sensor Networks (WSNs) and Internet of Things (IoT), we will measure their complexity by the number of exchanged messages between sensors (nodes or systems). Another specificity of sensor nodes is their energy consumption. Without going into details of the underlying protocol and for simplicity, we will suppose that the main consumption of these systems is due to the operations of sending and receiving messages. We will not consider the possibilities of sleeping, listening and interference consumptions. When a sensor node (the transmitter) sends a message to another sensor node (the receiver), a consumption, say of energy, at both the level of the transmitter and the level of the receiver arises. If we consider that each operation consumes 1 unit, then we can evaluate the complexity of a distributed algorithm applied to WSNs in terms of the number of Sending/Receiving (SR)-messages. Another important parameter that we have to take into account is the communication mode. A node of a WSN can send a message either in a direct mode (see Figure 4.2(a)) or in a broadcast mode (see Figure 4.2(b)). Each mode will result in a different consumption. In the direct mode, the consumption concerns the transmitter and the receiver and will be of two units for one message sending. However, in the broadcast mode, the consumption concerns the transmitter and all of its neighbors. If the transmitter has k neighbors, then the consumption will lead to $k+1$ units for one message sending. Hence, for one broadcast sending, the number of SR-messages equals $k+1$, and for n broadcast sendings, this number is $(k+1)n$, where k is the number of neighbors of each sender (or transmitter). Here, we have assumed that each transmitter has the same number of neighbors.

Based on this new parameter, the classical time complexity cannot give an accurate information on the energy consumption generated by the algorithm, since two algorithms having the same complexity will not consume the same energy. This can be further explained by an example. Two algorithms that have a time complexity of $O(kn^3 + n^2)$ and $O(n^3)$, respectively, can be considered as having the same complexity of $O(n^3)$. But in the case of WSNs, these algorithms will not lead to the same number of SR-messages. In a network with $n = 100$ nodes and $k = 2$ neighbors for each node the second algorithm requires 50% less SR-messages than the first.

The feasibility of communications in a WSN is another important parameter, because physical problems like collisions or interferences may arise when several nodes are sending messages at the same time to the same node. As a consequence, lost messages or retransmissions can lead to additional SR-messages. Such situations can be handled by adding delays to transmissions, but only at the expense of additional time. We will not consider these situations here. Another parameter specific to WSNs, which we will not take into account here, is the number of SR-messages of each node. It is possible that the complexity of algorithm 1 is smaller than that of algorithm 2, but algorithm 1 uses only a part of the network. In this case, some sensors or systems will be overstretched and in case they are autonomous, they will be dead quickly. Thus, even if other sensors survive, the network will become unusable.

4.4   Functions and message primitives

We will now introduce some functions and message primitives to be used by the algorithms presented in this chapter.

In Table 4.1 we define some functions and in Table 4.2 some message primitives necessary for the communication between sensor nodes. Note that the proposed algorithms require bidirectional communications.

Table 4.1: Functions of the proposed algorithms.

Function	Definition
getId()	returns the node identifier
delay(dt)	waits dt milliseconds before going to the next instruction
send(a,b)	sends the message a to the sensor node having the identifier b, or in a broadcast (if $b = *$)
read()	reads a received message. This function is blocking, which means that if there is no received message any more, it remains blocked in this instruction
read(wt)	waiting for receipt of messages. If there is no received message after wt milliseconds, then the execution will continue and go to the next instruction
getCurrentTime()	returns the local time of a node
add(v,t)	adds the value v at the top of the vector t
pop(t)	removes the value at the top of the vector t and returns it
getCoord()	returns the node coordinates (x, y)
getX()	returns the x-coordinate of the node
getY()	returns the y-coordinate of the node
getNumberOfNeighbors()	returns the number of neighbors of the node
stop()	stops the execution of the program

4.5   Trees and transmissions

4.5.1   Flooding and spanning tree

Flooding is a network routing algorithm, based on Breadth-First Search (BFS), in which every incoming message is sent through every outgoing link

Table 4.2: Message primitives and their definitions.

Primitive	Definition
T1	If you do not have a previous node, then I am that one for you (this type is used to determine the spanning tree)
T2	I acknowledge receipt (this process is used to determine if a node is a leaf or not)
T3	I send you the minimum value that I have received (this process is used to route the minimum of a branch to the root)
T4	I want to elect the global minimum (leader election)
AC	Ask for coordinates
CS	Send coordinates
SN	Select a node
DN	Reset message

except the one it arrived on [109]. Figure 4.3 illustrates how the execution of the flooding algorithm on an example can lead to a communication between nodes in form of a spanning tree of the network. The pseudo-code of the Flooding algorithm is given by Algorithm 10.

4.5.2 Flooding for Leaf Finding

If we want to determine the leaves of the resulting spanning tree, we may use the Flooding for Leaf Finding (FLF) algorithm. Its pseudo-code is given by Algorithm 11 which is based on the previously presented Flooding algorithm. The messages of type T1 are used for the flooding process and the messages of type T2 are used for acknowledgments. The algorithm works as follows: during the flooding process and after each transmitted T1 message (lines 5 and 6 for the root and lines 17 and 18 for the other nodes), the transmitter will wait for an acknowledgment, i.e., a T2 message (lines 12 and 20). At the beginning and with the exception of the root node (line 3), each node is considered as a leaf (line 8). If any transmitter receives an acknowledgment, then it will be considered as a non-leaf (lines 20 to 22).

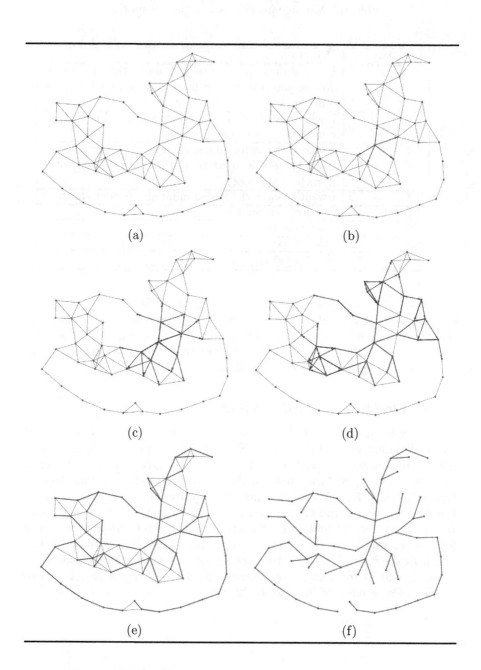

Figure 4.3: The Flooding algorithm: finding a spanning tree.

Algorithm 10: Flooding

Data: id_root

1   id = getId();
2   once = false;
3   **if** *(id==id_root)* **then**
4 | once = true;
5 | send(A, *);
6   **end**
7   **while** *(true)* **do**
8 | m = read();
9 | **if** *(m==A) and (once==false)* **then**
10 | | once = true;
11 | | send(A, *);
12 | **end**
13   **end**

4.6   Leader election

A leader in a distributed system is a node whose special task is to co-ordinate the execution of distributed activities in a most efficient way. It is an important concept widely used in distributed computing and a number of algorithms have been designed to solve the problem of leader election for various types of networks [27, 71, 94]. Such an election can be based on several values associated with a node: smallest or highest identifier, highest (remaining) local energy, etc. For our purposes, it will be convenient to first use the abstract criterion of a *value* associated with a node, to look for a node with minimum, maximum or optimum value, and later on to specify the term *value* by minimum or maximum x-coordinate which will be used as a starting point for our principal algorithms LPCN [76] and D-LPCN [100]. In this section, we present several algorithms for the search of such a point. We start with the Minimum Finding algorithm [99]. Next, we exhibit some algorithms that we have proposed in our previous work [21, 22, 23, 27]. These algorithms show an improvement of up to 90% over the MinFind algorithms, considering the energy consumption. The results of this comparison are given in Section 4.6.6.

4.6.1   Wait-Before-Starting

Let us assume that the node to elect is the one with minimum value (if there are several such nodes, we are free to choose one of them). We propose to use a concept that we call WBS (Wait-Before-Starting) [12, 22], which creates a relation between the value of a node x with a given waiting time unit gt.

Algorithm 11: Flooding for Leaf Finding (FLF)

    Data: id_root
    Result: leaf

1 id = getId();
2 **if** *(id==id_root)* **then**
3 | leaf = false;
4 | once = true;
5 | message = (T1, id);
6 | send(message, *);
7 **else**
8 | leaf = true;
9 | once = false;
10 **end**
11 **while** *(true)* **do**
12 | (type, r_id) = read() ;
13 | **if** *(type==T1) and (once==false)* **then**
14 | | once = true;
15 | | message = (T2, id);
16 | | send(message, r_id);
17 | | message = (T1, id);
18 | | send(message, *);
19 | **end**
20 | **if** *(type==T2)* **then**
21 | | leaf = false;
22 | **end**
23 **end**

Any node has to wait for a time $(t = x \times gt)$ before starting the election algorithm. The value of gt must be sufficiently high to allow termination of the process of informing all the nodes about an existing root. This option is useful if the first node, assumed to be the root, fails. If such a failure occurs, a second node, considered as the new root, will start the process and if both the first and the second nodes fail, a third node will start the process, and so on. The pseudo-code of this revision is given by Algorithm 12. The WBS algorithm works with integer values. If the leader election implies real values, then a variable transformation must be done before, and if the real values are very close, the WBS algorithm can be combined with other algorithms like BrOGO [21] which is low energy consuming and fault-tolerant.

4.6.2 Minimum Finding

So let us explain how we can elect a node with minimum value. For convenience, we call such a node a minimum. A local minimum, also called *Local*

Algorithm 12: Wait-Before-Starting (WBS)

    Data: x, gt

1   $once$ = false;
2   $t = x \times gt$;
3   **while** *(true)* **do**
4 | r_x = read(t);
5 | **if** *(r_x==null)* **then**
6 | | send(A, *);
7 | | stop();
8 | **end**
9 | **if** *(r_x==A and once==false)* **then**
10 | | $once$ = true;
11 | | send(A, *);
12 | | stop();
13 | **end**
14   **end**

Leader, is a node which has no neighbor with a value smaller than its own. Clearly, this value is not necessarily a global minimum.

4.6.2.1   Local Minima Finding

The Local Minima Finding (LMF) Algorithm 13 uses the same principle as the previously presented *MinFind* algorithm to determine if a node is a local minimum or not, with the exception that each node will send its coordinates only once, and after reception of messages from all its neighbors, it decides whether it is a local minimum or not in case it has received a smaller value than its own. The algorithm of finding local minima is given as follows:

Algorithm 13: Local Minima Finding (LMF)

    Data: wt, v
    Result: *local_min*

1   $local\_min$ = true;
2   $x_{min} = v$;
3   send(x_{min}, *);
4   **while** *(((x = read(wt))≠ null) and local_min)* **do**
5 | **if** *(x < x_{min})* **then**
6 | | $local\_min$ = false;
7 | **end**
8   **end**

4.6.2.2   Global Minimum Finding

In this section, we will present a distributed algorithm that allows to determine a leader representing the node with minimum value v. It is based on the Minimum Finding (MinFind) algorithm presented in [83, 99] which relies on the tree-based broadcast algorithm. The principle of this algorithm can be described as follows. First, each node of the network assigns its local value to the variable x_{min} assumed to represent the minimum value of the network (the leader). Then it will broadcast this value and wait for other x_{min} values coming from its neighbors. If a received value x_{min} is less than its local x_{min} value, then this one will be updated and broadcasted again. This process is repeated by each node as long as a received value is less than its local x_{min} value. After a certain time t_{max}, there will be only one sensor node that has not received a value that is smaller than its local x_{min} value. This node is the leader. The pseudo-code of this process is given by Algorithm 14, where t_0 is the time of the first execution of the algorithm (which can correspond to the first powering-on of a sensor node), t_c the current local time of a sensor node, and t_{max} the maximally tolerated running time of the algorithm from the first execution to the current time of a sensor node.

Algorithm 14: Minimum Finding (MinFind)

```
1  leader = true;
2  t₀ = getCurrentTime();
3  x_min = v;
4  send(x_min, *);
5  repeat
6  |    x = read();
7  |    if (x < x_min) then
8  |    |    leader = false;
9  |    |    x_min = x;
10 |    |    send(x_min, *);
11 |    end
12 |    t_c = getCurrentTime();
13 until (t_c - t₀ > t_max);
```

Evaluating the time complexity of this algorithm will help to set the value of t_{max}. To do this, let us consider a linear network with n nodes representing the worst case if, for example, we are searching the node with minimum x-coordinate. This node, situated on the extreme left, will send only 1 message and receive only 1 message. However, the node which is on the extreme right will receive $n-1$ messages and send $n-1$ messages, because it is the node with the largest x-coordinate. Moreover, each of the other nodes, except the one on the extreme left, has at least one node on its left. Therefore, these nodes will systematically send a message to broadcast the newly received x_{min} value.

Altogether, the message complexity is equal to $M[MinFind] = 2(n-1) = 2n - 2$ and if we consider that a sensor can send and receive messages at the same time (full-duplex communication), the time complexity is equal to $T[MinFind] = n-1$. This complexity is reduced in the case where the network is not linear and any node will receive d messages instead of $n-1$, where d represents the diameter[1] of the network and which is equal to n in the case of the linear network. Thus, the general message complexity of the MinFind algorithm is equal to $M[MinFind] = d - 1$. Since the time complexity is known, it is possible to estimate the value of t_{max}, the required time to find the leader. As an example, for a network with 100 nodes, where the size of each message is equal to 1024 bits, sampled with a frequency rate of 250 kb/s (case of the 802.15.4 standard) we need 409 ms to find the leader since we need 4.09 ms to send one bit. Using the CupCarbon simulator, we have simulated two networks with 100 sensor nodes each. The first one is linear (see Figure 4.4) and the second one is random (see Figure 4.1). The simulation results show that the leader is obtained in 409 $ms = 100 \times 4.09$ ms with a consumption of $1J$ to $9J$ per node for the linear network, and in 70 $ms = 18 \times 4.09$ ms with a consumption of $1J$ to $5J$ per node for the random network, the diameter of which is 17. Just note that in these calculations we do not take account of the time necessary for collision management.

Figure 4.4: A linear network with 100 sensor nodes.

4.6.3 Local Minima to Global Minimum (LOGO)

4.6.3.1 The concept

In the *MinFind* algorithm, each node is sending messages repeatedly to update its value each time the received value is smaller than its own. After a certain time, each node will be marked as a non-leader, except the leader which has the smallest value since this node will never receive a smaller value than its own. This process is time-consuming and requires a lot of broadcast messages, which makes it very energy consuming and impractical for real-world WSNs, because of possible collisions, for instance. To improve upon this, we propose a new approach in which each node will send a broadcast message once, in the aim to determine the local minima using the LMF algorithm (cf. Algorithm 13). Then each local minimum will send a message to a given reference node which will select the global minimum. This approach can be described as follows:

---

[1]The diameter d of a network is defined as the longest path of the shortest paths between any two nodes.

Step 1: Mark each node as a global minimum and select one node as a reference node (see Figure 4.5).

Step 2: Run the LMF algorithm to find the local minima and unmark the other nodes (see Figure 4.6).

Step 3: The reference node will send a message asking the local minima to send their values (see Figure 4.7).

Step 4: Each local minimum will send a message to the reference node and the reference node will determine the global minimum from the received local minima (see Figure 4.8).

Step 5: The reference node will send a message to the global minimum saying that it is the leader (see Figure 4.9).

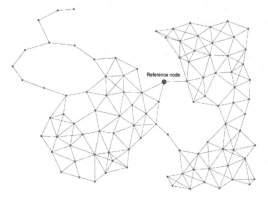

Figure 4.5: A network with a designated reference node.

4.6.3.2   The algorithm

To present the proposed algorithms, we observe that there are two algorithms. The first is executed by the reference node (Algorithm 15) and the second by the remaining nodes (Algorithm 16). Algorithm 15 takes as inputs a value x and the time wt required to select the global minimum. The output *global_min* is a variable which takes the value true if the current reference node is a leader (minimum) and false, otherwise. Algorithm 15 starts with an initialization (lines 1 to 3), and waits for 1 second (line 4), the necessary time to finish the process of determining the local minima. This time must be fixed according to the (possibly very important) number of neighbors of a sensor node. It is the time required for any sensor node to send and receive a message in a broadcast and receive mode. Then a T2 message is sent to ask the local minima to send their values x (lines 5 and 6). In line 8, the reference node will

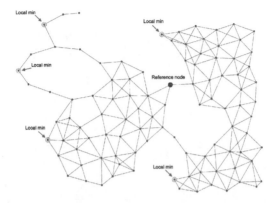

Figure 4.6: The local minima found by Algorithm 13.

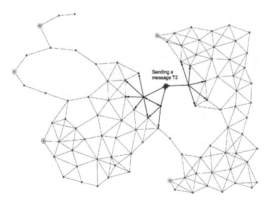

Figure 4.7: The reference node asks for local minima.

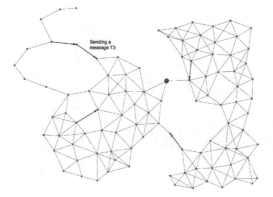

Figure 4.8: The local minima declare themselves to the reference node (gray arrows).

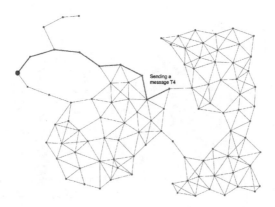

Figure 4.9: The reference node designates the global minimum (leader) and informs that node.

wait for the receipt of a message containing the id of the transmitter (r_id), the value r_x of the local minimum message and t, the stack of the path from the local minimum to the reference node. The receipt of a message within the next *wt* milliseconds means that a T3 message has been received by a local minimum. In this case, the received value r_x is tested whether it is smaller than the current value x_min which at the beginning was equal to the local value *x* (line 18). If this is the case, the reference node will be declared as a non-global minimum (line 19), the value of id_min will be updated with the value of r_id (line 20), the value of x_min will be updated with the value of r_x (line 21) and the route t from the reference node to the local minimum (id_min) will be assigned to t_min (line 22). Otherwise, if the received message is null (line 9), which can happen if the node does not receive any message during the *wt* milliseconds, then the reference node has received messages from all local minima. In this case and if the *global_min* value is equal to true, the reference node is the global minimum and the algorithm will stop (line 15). Otherwise, the route t is the one situated between the reference node and the global minimum. A T4 message will then be sent to the global minimum, containing the identifier id_min and using the route t (lines 11 to 13) in order to elect it as the leader.

Algorithm 16 for a non-reference node takes as input only the value *x*. The output *global_min* is a variable which takes the value true if the current node is a leader (global minimum) and false, otherwise. Each non-reference node starts with initializations (lines 1 to 5). The variable once1 is used to allow only once the reception of T2 messages and the variable once2 is used to accept only once any received T4 message. Then it starts the process of the LMF algorithm by sending in a broadcast a T1 message in order to test if it is a local minimum or not by comparing the values received from its neighbors with its own value *x*. If any received value is smaller than *x*, the node will be considered as a non-global minimum (lines 9 to 15). Once all the values of the neighbors received, the algorithm goes to the second step, where it will wait for a T2 message initiated by the reference node. In this case, it will route this message to its neighbors and if it is a local minimum (*global_min* = true) then it will send a T3 message as a response to the T2 message coming from the reference node, telling him that it is a local minimum (lines 22 and 23). Finally, it will be considered as a non-global minimum (line 24). The next part of the algorithm concerns the creation of the route from the local minimum to the reference node. If any node receives a T3 message, then it will add itself to a stack t (line 28) representing the route from the local minimum to the reference node, and route it again to the node p_id which had sent him previously a T2 message (lines 29 and 30). As soon as all the non-reference nodes have done this step, the reference node will be in the situation to have received all the routes and values from the local minima and, thus, to choose the global minimum. It will then send a T4 message to elect the global minimum (lines 11 to 13 of Algorithm 15). Finally, each non-reference node, which receives a T4 message (line 32) containing the route t and the

---

Algorithm 15: LOGO for the reference node

Data: x, wt

Result: *global_min*

1 id = getId();

2 x_min = x;

3 *global_min* = true;

4 delay(1000);

5 message = (T2, id, null, null);

6 send(message, *);

7 **while** *(true)* **do**

8 $\quad$ (type, r_id, r_x, t)=read(wt);

9 $\quad$ **if** *(type==null)* **then**

10 $\quad\quad$ **if** *(global_min==false)* **then**

11 $\quad\quad\quad$ n_id = pop(t_min);

12 $\quad\quad\quad$ message = (T4, id_min, null, t_min);

13 $\quad\quad\quad$ send(message, n_id);

14 $\quad\quad$ **else**

15 $\quad\quad\quad$ stop();

16 $\quad\quad$ **end**

17 $\quad$ **else**

18 $\quad\quad$ **if** *((type==T3) and (r_x < x_min))* **then**

19 $\quad\quad\quad$ *global_min* = false;

20 $\quad\quad\quad$ id_min = r_id;

21 $\quad\quad\quad$ x_min = r_x;

22 $\quad\quad\quad$ t_min = t;

23 $\quad\quad$ **end**

24 $\quad$ **end**

25 **end**

---

identifier r_id of the leader, will test if its identifier id matches the received identifier r_id (line 35). If so, it will be elected (lines 36 and 37). Otherwise, it will route the same message to the next sensor node having the identifier n_id pulled from the route t (lines 39 to 41).

4.6.4   Branch Optima to Global Optimum (BrOGO)

4.6.4.1   The concept

To overcome the limitations imposed by the *MinFind* algorithm and already outlined at the beginning of Section 4.6.2.2, we propose a second approach in which a given reference node will start the FLF algorithm presented in Section 4.5.2 for the determination of a spanning tree of the network. Then each leaf of this tree will route a message from itself to the reference node

---

Algorithm 16: LOGO for a non-reference node

Data: x

Result: *global_min*

1   id = getId();

2   x_min = x;

3   *global_min* = true;

4   once1 = false;

5   once2 = false;

6   message = (T1, id, x_min, null);

7   send(message,*);

8   **while** *(true)* **do**

9   (type, r_id, r_x, t) = read();

10   **if** *(type == T1)* **then**

11   **if** *(r_x < x_min)* **then**

12   x_min = r_x;

13   *global_min* = false;

14   **end**

15   **end**

16   **if** *((type == T2) and (once1 == false))* **then**

17   once1 = true;

18   p_id = r_id;

19   message = (T2, id, null, null);

20   send(message, *);

21   **if** *(global_min == true)* **then**

22   add(id, t);

23   message = (T3, r_id, x_min, t);

24   send(message, p_id);

25   *global_min* = false;

26   **end**

27   **end**

28   **if** *(type == T3)* **then**

29   add(id, t);

30   message = (T3, r_id, r_x, t);

31   send(message, p_id);

32   **end**

33   **if** *((type == T4) and (once2 == false))* **then**

34   once2 = true;

35   **if** *(r_id == id)* **then**

36   *global_min* = true;

37   stop();

38   **else**

39   n_id = pop(t);

40   message = (T4, r_id, null, t);

41   send(message, n_id);

42   **end**

43   **end**

44   **end**

(the root). During the routing process, the minimum value will be sent to the reference node in order to determine the branch leader (branch minimum). That is to say, in each hop, the receiver node will check if the received value is smaller than its own, in which case the routed value will be the received one, otherwise, it routes its own value. Upon receipt of the minima from all the branches, the reference node will select the global minimum and send a message to the corresponding node to elect it. The main steps of the BrOGO algorithm are given as follows:

Step 1: Let us consider the network of Figure 4.10(a). We run the FLF algorithm to determine a spanning tree and its leaves (or branches) (see Figure 4.10(b)).

Step 2: Each leaf will route a message from itself to the root (reference node). This process will allow to route the minimum value in each branch situated between the reference node and each leaf. The reference node will determine the global minimum from the received values (see Figure 4.10(c)).

Step 3: The reference node will send a message to the global minimum saying that it is the global minimum (see Figure 4.10(d)). The result of the leader election for this example is shown by Figures 4.10(e) and (f).

4.6.4.2   The algorithm

To present the proposed algorithms, we note that there are, just as for LOGO, two algorithms. The first one is executed by the reference node (Algorithm 17) and the second one by the remaining nodes (Algorithm 18). Algorithm 17 of the reference node takes as inputs a value x and a time wt required before selecting the leader using a T4 message. The output *leader* is a variable which takes the value true if the reference node is a leader (minimum) and false, otherwise. It starts with an initialization (lines 1 to 4) and sends a T1 message with its identifier id (line 4) in order to start the process of finding a spanning tree by use of the flooding routing protocol. This process allows to determine for each node its predecessor (p_id) necessary to route information from each leaf to the reference node (branch routing process).

Once the spanning tree is determined, the reference node will wait for the reception of T3 messages from each branch (line 8). These messages will route the minimum value from each leaf to its neighbors (nodes having the identifier r_id), to the neighbors of these neighbors and so on, until to arrive at the reference node which will select the minimum from all the received branch minima (lines 9 to 13). If there is no received message within a time period wt, all the branch minima have been received, and the minimum can be determined. The current reference node will then elect the leader by informing it using a T4 message (lines 15 to 18).

Algorithm 18 for the remaining nodes takes as input values x and wt.

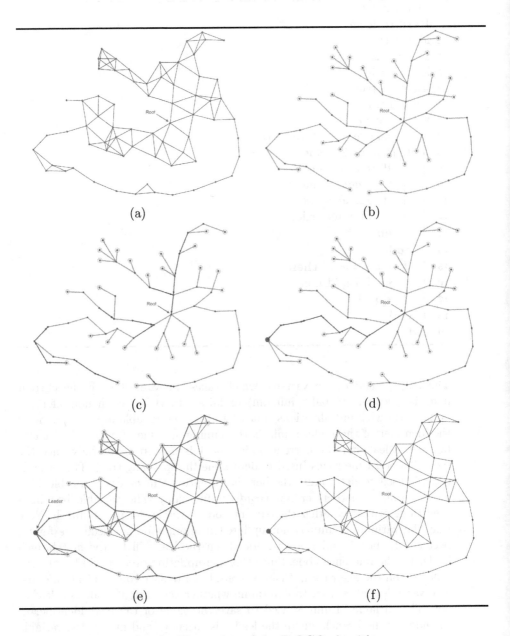

Figure 4.10: An illustration of the BrOGO algorithm.

Algorithm 17: BrOGO for the reference node

Data: x, wt

Result: *leader*

```
1  id = getId();
2  x_min = x;
3  id_min = id;
4  leader = true;
5  send((T1, id), *);
6  while (true) do
7  |    (type, r_id, r_x) = read(wt);
8  |    if (type==T3) then
9  |    |    if (r_x<x_min) then
10 |    |    |    leader = false;
11 |    |    |    x_min = r_x;
12 |    |    |    id_min = r_id;
13 |    |    end
14 |    end
15 |    if (type==null) then
16 |    |    send(T4, id_min);
17 |    |    stop();
18 |    end
19 end
```

The output *leader* is a variable which takes the value true if the current node is a leader (global minimum) or false, otherwise. Each non-reference node starts with initializations (lines 1 to 9). The variable id corresponds to the identifier of the node, x_min is the minimum value of a branch situated between a leaf and the current node, once1 is used to allow only once the reception of T1 messages (used to determine the spanning tree). The variable twait is used to determine whether the reception will be done in a blocking mode (*read()* function) or not (*read(wt)* function). The variable br_min is used to determine whether the current node is the leader in its branch. Note that the variable br_min is not updated if another branch leader is found, because the last found branch leader from the leaf will be the first branch leader for the reference node. Since this non-updating process will be without influence on the algorithm, it can be considered to save communication costs. The variable leaf is used to determine whether the current node is a leaf or not. The variable id_min is used to save the value of the next identifier of the node of the branch where the leader is situated, and finally, the variable *leader* is used to determine whether the current node is the leader or not. Lines 10 to 14 are used to read the received messages. Lines 15 to 18 are used to detect whether the current node is a leaf and if so, it will be considered as a branch leader (line 16). Then it will send its value x to its neighbor p_id.

This value will be forwarded to the root if it is the minimum one, otherwise, another smallest value will be forwarded and the corresponding node will be declared as a branch leader. Finally, any branch leader will wait for a T4 message in order to be elected as the network leader (global minimum). If the current node is not a leaf (line 19), and if it receives the T1 message for the first time, this means that the process of determining a spanning tree is running. The current node will then send a T1 message to its neighbors to continue this process (line 24) as well as an acknowledgment (T2) message to its previous neighbor. The objective of an acknowledgment message is to inform the neighbor that it is not a leaf. The variable p_id of line 21 will be used to route received messages to the reference node. Lines 26 to 29 are executed if a T2 message is received, which means that the current node is not a leaf (line 27). At the same time, the variable n_id (next id) will be assigned. This variable will be used to route the T4 message from the reference node (root). Lines 30 to 38 are executed if a T3 message is received. This means that the process of routing the branch minima is running. In line 31, a test is used to determine which subbranch contains the minimum. If a received value r_x is smaller than the current smallest value x_min (line 31), it will be updated by that one (line 32) and the identifier of the corresponding node (line 33). Therefore, since the received value is smaller than its own, the current node cannot be a leader of the current branch (line 34). In line 36, twait is set to true in order to wait for T4 messages in case that the current branch contains the leader. Line 37 is used to route the branch leader to the root by sending a T3 message to the previous neighbor. Finally, lines 39 to 47 are used to elect the leader. This phase is started by the reference node (line 16 of Algorithm 17), which will send a T4 message only through the branch containing the leader. In other words, if a node in this branch receives a T4 message and if it is a branch leader (line 40) then it will be elected as the leader (global minimum). Otherwise, it will send another T4 message to its next neighbor until the leader is reached.

Algorithm 18: BrOGO for a non-reference node

    Data: x, wt
    Result: *leader*

```
 1  id = getId();
 2  x_min = x;
 3  once1 = false;
 4  twait = false;
 5  br_min = true;
 6  leaf = true;
 7  id_min = 0;
 8  leader = false;
 9  while (true) do
10      if (twait==false) then
11      |   (type, r_id, r_x) = read(wt);
12      else
13      |   (type, r_id, r_x) = read();
14      end
15      if ((type==null) and (leaf==true)) then
16      |   br_min = true;
17      |   twait = true;
18      |   send((T3, id, x), p_id);
19      else
20      |   if ((type==T1) and (once1==false)) then
21      |   |   p_id = r_id;
22      |   |   once1 = true;
23      |   |   send((T2, id), r_id);
24      |   |   send((T1, id), *);
25      |   end
26      |   if (type==T2) then
27      |   |   leaf = false;
28      |   |   n_id = r_id;
29      |   end
30      |   if (type==T3) then
31      |   |   if (r_x<x_min) then
32      |   |   |   x_min = r_x;
33      |   |   |   id_min = r_id;
34      |   |   |   br_min = false;
35      |   |   end
36      |   |   twait = true;
37      |   |   send((T3, id, x_min), p_id);
38      |   end
39      |   if (type==T4) then
40      |   |   if (br_min==true) then
41      |   |   |   leader = true;
42      |   |   else
43      |   |   |   send(T4, id_min);
44      |   |   end
45      |   end
46      end
47  end
```

4.6.5 Dominating Tree Routing (DoTRo)

4.6.5.1 The concept

The Dominating Tree Routing (DoTRo) algorithm is based on a tree routing protocol. It starts from local leaders which will run, as a root, the process of flooding [109] to determine a spanning tree. During this process the value of the leader (root) will be routed. If two spanning trees meet each other, then the tree routing the better value will continue its process while the other one will stop. Based on the example of Figure 4.11, we are now going to present the main steps of the DoTRo algorithm, where we assume that the leader is the node having the maximum value:

Step 1: For the network of Figure 4.11(a), we run the LMF algorithm (cf. Algorithm 13) to determine the local minima. The obtained result is shown by Figure 4.11(b) where we have two local minima: 1 and 4 because they are the only nodes that do not have any neighbor with a value smaller than their own.

Step 2: Each local minimum will start the flooding process to route the leader's value (local minimum) over the tree (see Figure 4.11(c)).

Step 3: When two trees meet, as is the case of the center node in Figure 4.11(c), the light gray tree chooses the center node with value 1 and the dark gray tree will choose the same center node with value 4. Since 1 is less than 4, the center branch of the dark gray tree will stop the flooding process, whereas the other branch will continue, and the center branch of the light gray tree will continue the flooding process. The obtained result is shown by Figure 4.11(d). Figures 4.11(e) and (g) show another meeting of two trees, and Figures 4.11(f) and (h) show the result of the DoTRo algorithm after this meeting.

Step 4: Each local minimum will wait for a given time, sufficient to finish the process of flooding. If after this time there is no received message anymore, the corresponding local minimum will become the leader.

4.6.5.2 The algorithm

The pseudo-code of the DoTRo algorithm is given by Algorithm 19. It is composed of three main parts: (1) the initialization (lines 1 to 4) where each node is considered as a leader (line 4), (2) the LMF algorithm (line 5 and lines 8 to 18) and (3) the flooding process (lines 21 to 30). The second part is explained above and in the third part, each node will wait for a message representing the value of the local minimum routed during the flooding process. If the received value is less than its own value, it will be considered as a non-leader node and route the received value to continue the flooding process. Otherwise, the node will do nothing, which will stop the process of flooding.

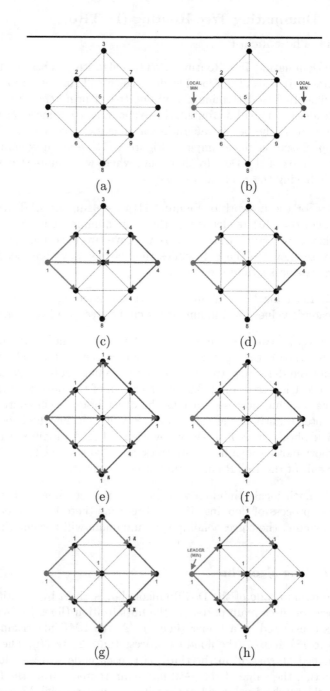

Figure 4.11: An illustration of the DoTRo algorithm.

Algorithm 19: Dominating Tree Routing (DoTRo)

Data: x, wt_1, wt_2
Result: *leader*

1   $id = \text{getId}()$;
2   $x_{min} = x$;
3   $step = 1$;
4   $leader = \text{true}$;
5   $\text{send}(x, *)$;
6   **while** *(true)* **do**
7   **if** *(step == 1)* **then**
8         $rx = \text{read}(wt_1)$;
9   **if** *(rx == null)* **then**
10            $step = 2$;
11   **if** *(leader == true)* **then**
12               $\text{send}(x, *)$;
13   **end**
14   **else**
15   **if** *(rx < x)* **then**
16               $leader = \text{false}$;
17   **end**
18   **end**
19   **end**
20   **if** *(step == 2)* **then**
21         $rx = \text{read}(wt_2)$;
22   **if** *(rx == null)* **then**
23            $\text{stop}()$;
24   **else**
25   **if** *(rx < x_{min})* **then**
26               $leader = \text{false}$;
27               $x_{min} = rx$;
28               $\text{send}(rx, *)$;
29   **end**
30   **end**
31   **end**
32 **end**

4.6.6   Comparison of the leader election algorithms

In this section, we will compare the 5 leader election algorithms presented above: MinFind, LOGO, BrOGO, DoTRo and WBS. The comparison will be based on the total number of sent and received (SR) messages metric. We can also use the term "consumption" to specify this metric. To do this comparison and for more clarity, we have considered 12 situations as illustrated by the star graph of Figure 4.12, and we will consider as a leader the node with the smallest x-coordinate. These situations can be explained in more detail as follows:

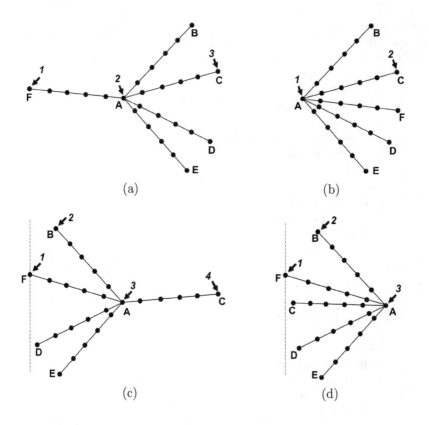

Figure 4.12: Configurations of a star graph with different reference nodes.

1. Graph (a)-1: is a star graph centered around node A, with 4 branches of 5 nodes on its right side, 1 branch of 5 nodes on its left side and where the reference node is the leader represented by node F having the smallest x-coordinate (Figure 4.12(a)).

2. Graph (a)-2: is the same as Graph (a)-1, where the reference node is node A, the center of the star with 5 branches of 5 nodes each (Figure 4.12(a)).

3. Graph (a)-3: is the same as Graph (a)-1, where the reference node is node C, a leaf of the star which is not the leader (Figure 4.12(a)). This graph can be considered as a tree rooted at C with 4 branches of 10 nodes, and where the leader is one of its leaves (here node F).

4. Graph (b)-1: is a star graph centered around node A, with 5 branches of 5 nodes on its right side only, and where the reference node is the node represented by number 1, the leader having the smallest x-coordinate (Figure 4.12(b)).

5. Graph (b)-2: is the same as Graph (b)-1, where the reference node is C, a leaf of the star which is not the leader (Figure 4.12(b)). As in Situation 3, this graph can be considered as a tree rooted at C with 4 branches of 10 nodes, but in this case, the leader is not a leaf (here node A).

6. Graph (c)-1: is a star graph centered around node A, with 4 branches of 5 nodes on its left side, 1 branch of 5 nodes on its right side and where the reference node is the leader F, a leaf of the star having the smallest x-coordinate (Figure 4.12(c)).

7. Graph (c)-2: is the same as Graph (c)-1, where the reference node is B, which is a local minimum but not the leader (Figure 4.12(c)).

8. Graph (c)-3: is the same as Graph (c)-1, where the reference node is the node represented by 3, the center of the star (Figure 4.12(c)). This situation differs from the situation 2 (Graph (a)-2) on the number of local minima, where we have 3 additional local minima (B, D and E).

9. Graph (c)-4: is the same as Graph (c)-1, where the reference node is the node represented by 3, a leaf of the star which is not the leader and situated on the right side of the star (Figure 4.12(c)).

10. Graph (d)-1: is a star graph centered around node A, with 5 branches of 5 nodes on only its left side, and where the reference node is the leader represented by F, a leaf of the star having the smallest x-coordinate (Figure 4.12(d)).

11. Graph (d)-2: is the same as Graph (d)-1, where the reference node is B, a leaf of the star which is a local minimum but not the leader (Figure 4.12(d)).

12. Graph (d)-3: is the same as Graph (d)-1, where the reference node is A, the center of the star (Figure 4.12(d)).

These situations are chosen so that to obtain different configurations in terms of the following parameters:

- The number of local minima,

- The number of branches of the tree starting from the reference node (root),

- The depth of the branches,

- The depth of the leader from the root,

- The type of the root: Leader (L), Local Minimum (M) or Nothing (N).

Table 4.3 shows the parameters corresponding to each of the 12 situations presented above.

Table 4.3: The parameters of configurations of the graphs used to compare the presented algorithms.

Graph	(a)			(b)		(c)				(d)		
	1	2	3	1	2	1	2	3	4	1	2	3
Number of local minima	1	1	1	1	1	4	4	4	4	5	5	5
Number of branches	4	5	4	5	4	4	4	5	4	4	4	5
Depth of the branches	10	5	10	5	10	10	10	5	10	10	10	5
Depth of the leader	0	5	10	0	5	0	10	5	10	0	10	5
Type of the root	L	N	N	L	N	L	M	N	N	L	M	N

We have executed the 5 presented algorithms on each of these graphs and obtained the histograms of Figure 4.13. The corresponding values are given in Table 4.4.

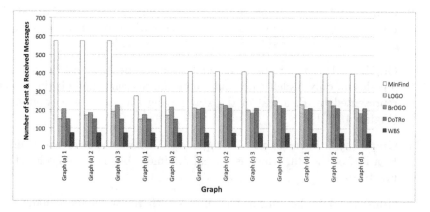

Figure 4.13: Histograms of the total number of sent and received messages given by different graph topologies.

As we can see, the MinFind algorithm requires a very important number of SR messages. Figure 4.14 and Table 4.5 show that these numbers have

Table 4.4: Total number of sent and received messages given by different graph topologies.

	Graph (a)			Graph (b)		Graph (c)				Graph (d)		
	1	2	3	1	2	1	2	3	4	1	2	3
MinFind	576	576	576	276	276	409	409	409	409	398	398	398
LOGO	152	172	192	152	172	212	232	202	252	232	252	212
BrOGO	206	186	226	176	216	206	226	186	226	206	226	186
DoTRo	152	152	152	152	152	212	212	212	212	212	212	212
WBS	76	76	76	76	76	76	76	76	76	76	76	76

overheads ratios from 22% to 87%. We can also see that in Graphs (a) and (b), LOGO requires less SR messages than BrOGO, and in Graphs (c) and (d), BrOGO requires less SR messages than LOGO. We can conclude from this that the higher the number of local minima in the graph, the more SR messages are required by LOGO than by BrOGO.

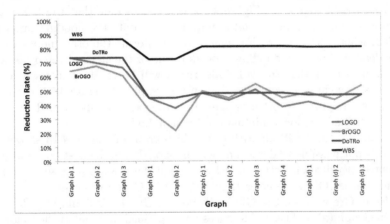

Figure 4.14: Reduction rate between the presented algorithms and the classical MinFind algorithm.

In [21, 23, 26, 27], it has been shown that these ratios can exceed 94% in the case of large random graphs.

The WBS algorithm can be used only if the considered values are integer. It cannot be used, for example, to find the node having the smallest x-coordinate since the GPS coordinates are real values. In spite of this limitation, the WBS algorithm can be useful or even essential to improve the other algorithms, where the reference node starting the flooding process is determined manually. In real applications, if this node is broken, the process of the leader election cannot be performed and, even worse, in the case where this node is required to start the boundary node finding algorithm or any other algorithm, the process will not be started at all. Since the WBS algorithm is fault-tolerant, it will find

Table 4.5: Reduction rate values between the presented algorithms and the classical MinFind algorithm.

	Graph (a)			Graph (b)		Graph (c)				Graph (d)		
	1	2	3	1	2	1	2	3	4	1	2	3
LOGO (%)	74	70	67	45	38	48	43	51	38	42	37	47
BrOGO (%)	64	68	61	36	22	50	45	55	45	48	43	53
DoTRo (%)	74	74	74	45	45	48	48	48	48	47	47	47
WBS (%)	87	87	87	72	72	81	81	81	81	81	81	81

the leader even when the node with smallest value is broken because it allows to find the leader among the nodes that are operational. In the following, we will use the WBS algorithm to find the node having the minimum id instead of the minimum x-coordinate. To be able to compare it with the other algorithms, we assume that the node with minimum id is also the one with minimum x-coordinate.

From the histogram of Figure 4.13 we can conclude that the WBS algorithm gives better results in all situations since only the flooding process is needed, where the node with smallest value informs the other nodes that it is the leader. Since the flooding process is one of the steps of the other algorithms, it is clear that the WBS algorithm will be the best in terms of SR messages. We can also conclude that in the graphs of Figure 4.12, the flooding process requires 76 SR messages. It is the same number of SR messages required to find the local minima (cf. Section 4.6.2).

Excluding the WBS algorithm, the histogram of Figure 4.13 shows also that it is not possible to determine which algorithm is the best one since each presented algorithm is better in at least one of the situations presented above. For instance, LOGO is better than BrOGO in Graph (a)-1 and better than DoTRo in Graph (c)-3, BrOGO is the best in Graph (d)-3 and DoTRo is the best in Graph (a)-3. These differences are explained in detail in the following. Note that the LOGO algorithm is never better than the other algorithms at the same time.

In Graph (a)-1, where the reference node is F, LOGO and DoTRo algorithms give the same results since both of them start from Local Minima Finding (cf. Section 4.6.2), which will find only one local minimum. The next step in each algorithm is flooding. In LOGO, the reference node F will send a message to ask each local minimum to send it its value. Since only the reference node is the local minimum, there is no received message in this step, and since the reference node is the leader, there is no message to send to the leader to elect it. In DoTRo, each local minimum will start the flooding process to inform the others that it is the leader. Since node F is the only local minimum, only one flooding process will be started. This justifies why these two algorithms require the same number of SR messages. BrOGO requires more messages because from the node F we can determine 4 branches of 10 nodes each, which will lead to a reception of 4 messages from each branch. This requires $(4 \times 20 = 80)$ SR messages.

In Graph (a)-2, the reference node is A with the exception of DoTRo, where the reference node is determined by the Local Minima Finding algorithm which will always determine the same node F as in the previous case. That is why the consumption remains the same. The LOGO algorithm requires more messages ($+20$ SR messages), because node A will receive a message from node F (5 sendings + 5 receptions $=$ 10 SR messages) since F is a local minimum and the reference node A has to inform node F that it is the leader (5 sendings + 5 receptions $=$ 10 SR messages). However, the BoROGO algorithm requires less messages because even there is one more branch (5 branches of 5 nodes) compared with the previous case (4 branches of 10 nodes), in this case, the branches are smaller. This will lead to $5 \times 5 \times 2 = 50$ SR messages to send the branch minima to the reference node A. In the previous case, we have 4 branches of 10 nodes between each leaf to the previous reference node F, which requires $4 \times 10 \times 2 = 80$ SR messages to send the branch minima to the reference node A.

In Graph (a)-3, the reference node is C with the exception of DoTRo, as explained in the previous case. The LOGO algorithm requires more messages than in the previous cases because the path from the local minimum to the reference node C is longer (5 additional nodes with respect to the previous case). Then there will be $5 \times 2 = 10$ additional SR messages from the local minimum node F to the reference node C to send its value and $5 \times 2 = 10$ additional SR messages from the reference node C to the local minimum F to elect it as a leader. BrOGO requires more messages, as well ($+20$ SR messages) for the same reason if we compare with the situation of Graph (a)-1.

In Graph (b)-1, we move the left branch (A,F) of Graph (a) to the right side of the center and the reference node A. In this case, the consumption of DoTRo remains always the same for the reasons explained above. LOGO requires the same number of SR messages as in the case of Graph (a)-1. BrOGO requires less SR messages than in the case of Graph (a)-2 (-10 SR messages) because there is no message to send to the leader, since the reference node and the leader are the same node A.

In Graph (b)-2, the reference node is C with the exception of DoTRo, as explained above. The consumption of LOGO is the same as in the case of Graph (a)-2 because the local minimum A will send a message to the reference node C to send its value and this last node will send a message to the node A to elect it. BrOGO has the same configuration as in Graph (a)-3 except the message of leader election which is routed to the leader with less messages (10 SR messages instead of 20).

The Graphs (c) and (d) have more local minima than previously presented graphs. The consumption of LOGO and DoTRo depends on the number of the local minima. Thus, they require more SR messages to elect the leader.

In Graph (c)-1, LOGO requires more SR messages than in all the previous cases. If we compare with Graph (a)-1, we find that there is an overhead of 60 SR messages. This is justified by the 3 additional local minima (B, D and E) where each of them requires 20 SR messages to send its value to the reference node F. BrOGO algorithm in this graph has the same configuration as in Graph (a)-1. That is why we obtain the same result. In the same way as LOGO, DoTRo requires more SR messages compared with the previous cases because of the additional number of local minima.

In Graph (c)-2, each of LOGO and BrOGO requires 20 additional SR messages compared with the previous case so that the reference node B sends the election message to the leader node F.

In Graph (c)-3, the LOGO algorithm requires less SR messages to send each local minimum to the reference node A if we compare with Graph (c)-1. There are 4 local minima and each of them requires 10 SR messages to send its value to the reference node A and 10 SR messages for this reference node to elect the leader F. Then we obtain a total number of 50 SR messages for sending the local minima values and the election. This represents 10 SR messages less than in Graph (c)-1. In this case, except the reference node F, there are 3 local minima that require 20 SR messages for each one to send its value to the reference node. This leads to a total number of 60 SR messages. BrOGO requires also 20 SR messages less than in the case of Graph (c)-1. This can be explained in the same way as for LOGO. The number of the branches from node F is equal to 4. Each of them requires 20 SR messages to send its minimum to the reference node, which leads to a total of 80 SR messages. In case where the node A is the reference node, we obtain 5 branches of 5 nodes. Each branch requires 10 SR messages to send its minimum to the reference node A. In addition, 10 SR messages are required to elect the leader node F. This leads to a total of 60 SR messages, which represents a reduction of 20 messages with respect to the case of Graph (c)-1.

In Graph (c)-4, LOGO requires more SR messages because there are 4 local minima. It has the same situation as the Graph (c)-2 with 1 additional local minimum which requires 20 additional SR messages (10 SR messages to send its value to the reference node C and 20 SR messages to elect the leader F). BrOGO has exactly the same situation as the Graph (c)-2.

Graph (d) looks like Graph (c) where the branch (A,4) is moved to the left of the center A. The situation remains the same for DoTRo in all the three situations of Graph (d).

In Graph (d)-1, LOGO has the same situation as Graph (c)-2, and BrOGO has the same situation as Graph (c)-1.

In Graph (d)-2, LOGO as well as BrOGO have the same situation as Graph (c)-4.

In Graph (d)-3, LOGO has the same situation of Graph (c)-1, and BrOGO has the same situation as Graph (c)-3.

Finally, we can conclude from these situations and from Figure 4.14 that if the considered values are integer and the nodes of the graph are synchronized, the best algorithm to use is WBS. Otherwise, in most of the situations, the DoTRo algorithm gives the best results in terms of SR messages. The superiority of the DoTRo algorithm over the others can be shown for the case of a non-connected graph composed by many connected components as shown by Figure 4.15. Since the DoTRo algorithm starts from the Local Minima Finding algorithm, this will launch the DoTRo algorithm from each local minimum of each connected component of the graph. This will lead to find the global minimum of each connected component as represented by the 4 dark gray nodes pointed to by the arrows in Figure 4.15. On the other hand, if we execute the LOGO or the BrOGO algorithm, we will obtain only one global minimum of one connected component. This minimum depends on the reference node and the global minimum to be found is the one belonging to the connected component containing the reference node.

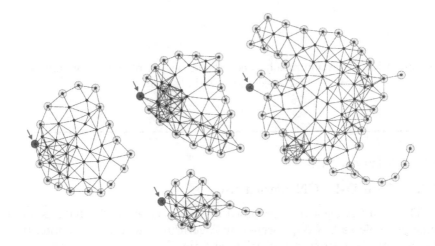

Figure 4.15: A disconnected Euclidean graph with several global minima.

It is possible to improve this situation for LOGO as well as for BrOGO by fixing automatically the reference node using WBS, just as in R-LOGO [23] presented in Section 5.5.10 and as in R-BrOGO presented in Section 5.5.11. As shown by Figure 4.16, this will lead to the same result as the one obtained by DoTRO but with additional SR messages generated by WBS. Note that in the case of Graph (b)-2, BrOGO requires more SR messages than the MinFind algorithm with an increasing ratio of 6%.

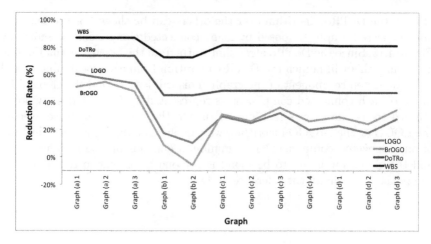

Figure 4.16: Reduction rate between the presented revised algorithms and the classical MinFind algorithm.

4.7   Polygon hull

4.7.1   The D-LPCN algorithm

The D-LPCN algorithm, proposed by Bounceur et al. [28, 100], uses the same principle as LPCN presented by Algorithm 5 with the exception that the program is written in a distributed form. This means that each sensor node will execute its own program, and by communicating with its neighbors, it can decide whether it is a boundary node or not. Algorithm 20 shows the main steps of the D-LPCN algorithm that can be described as follows:

Step 1 (lines 1 to 4): Initialization.

Step 2 (line 5): Run Algorithm 2 to determine whether the current node is a starting node or not.

Step 3 (lines 6 to 10): If the current node is a starting node, then it launches the D-LPCN algorithm by broadcasting the "AC" message to its neighbors in order to ask for their coordinates.

Step 4 (lines 12 to 13): Each node is waiting to receive a message.

Step 5 (lines 14 to 17): The variable i determines the number of received "CS" messages. Receiving n "CS" messages means that the coordinates of all neighbors have been received.

Step 6 (lines 18 to 20): If the node receives an "AC" message, then it will send a "CS" message containing its coordinates to the transmitter having the identifier *id*.

Step 7 (lines 21 to 27): For all received "CS" messages, the node calculates the minimum angle formed with them by taking into account the intersections with the received boundary set (in line 15). This calculation is done using the function angleWI (angle without intersections).

Step 8 (lines 28 to 35): The reception of an "SN" message by a node means that this node has been selected as a boundary node by its previous boundary neighbor node. This node will then restart the process of finding the next boundary node by broadcasting a "CS" message.

Finally, Algorithm 20 stops when the first boundary node is selected a second time with an "SN" message (lines 29 and 30).

To better explain how this algorithm works, we will use the example of Figure 4.17 which represents a WSN with eight sensor nodes. We consider the set $S = \{S1, S2, ..., S8\}$ of these nodes and the set $\mathbb{B}$ of boundary nodes, which initially is empty.

First and after the initialization (*Step 1*), we run Algorithm 14 (*Step 2*). The only node which will be considered as the starting node is $S1$, and thus the boundary set is updated to *boundary_set* $= \{S1\}$.

Next, the node $S1$ broadcasts an "AC" message to its neighbors $N(S1) = \{S2, S3, S4, S7\}$ (*Step 3*) to ask for their coordinates (see Figure 4.17(a)) while the other nodes are waiting for the receipt of messages (*Step 4*).

Next, each neighbor node $S2, S3, S4$ and $S7$ sends a "CS" message (*Steps 5 and 6*) to the boundary node $S1$ (see Figure 4.17(b)) which will calculate the angle formed by the fictive node $S1'$ with itself and with each of its neighbor nodes $S2, S3, S4$ and $S7$ (*Step 7*). This situation is illustrated by Figure 4.17(c).

Then, as shown by Figure 4.17(d), the boundary node $S1$ will send an "SN" message to the new boundary node $S3$ which will update its boundary set to *boundary_set* $= \{S1\}$ (*Step 8*). The obtained situation is shown by Figure 4.17(e).

The next boundary node $S3$ will then perform the same procedure from *Step 2* on. Figure 4.17(f) shows the final iteration of the D-LPCN algorithm.

Altogether, the pseudo-code of D-LPCN is given by Algorithm 20.

4.7.2 The D-RRLPCN algorithm

The D-RRLPCN algorithm [28, 100] uses the same principle as RRLPCN presented by Algorithm 9 with the exception that it is written in a distributed form. Figure 4.19 shows an execution example of the RRLPCN on a graph of 10 nodes. This algorithm starts from the dark gray node (see Figure 4.19(a)) the D-LPCN algorithm where it will determine the next node among its neighbors. The next node is the neighbor which forms the minimum polar angle between

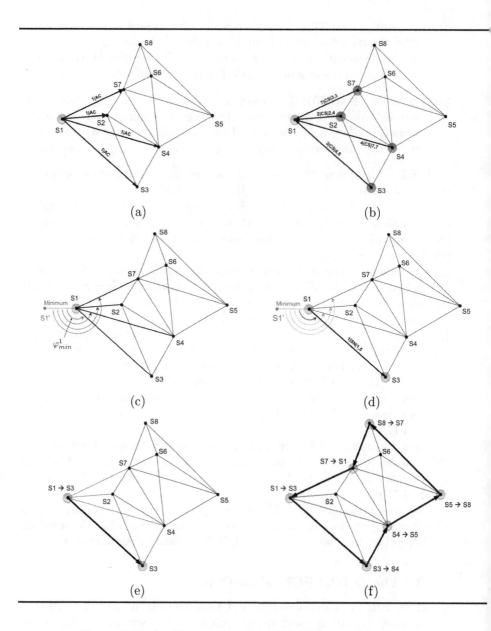

Figure 4.17: An illustration of the D-LPCN algorithm.

Algorithm 20: D-LPCN

1 boundary = false; phi_min = 10; phi_max = -10;
2 c_id = getId(); c_coord = getCoord();
3 boundary_set = ∅;
4 i=0; n = getNumberOfNeighbors(); selected = false;
5 Run any leader election algorithm to determine the value of first_node;
6 **if** *(first_node)* **then**
7   boundary = true;
8   p_coord = (c_coord.x−1, c_coord.y);
9   send(c_id+"|"+"AC", *);
10 **end**
11 **repeat**
12   id = read();
13   type = read();
14   **if** *(i==n)* **then**
15   boundary_set = boundary_set ∪ {c_id};
16   send(c_id+"|"+"SN"+"|"+c_coord+"|"+boundary_set, n_id);
17   **end**
18   **if** *(type=="AC")* **then**
19   send(c_id+"|"+"CS"+"|"+c_coord, id);
20   **end**
21   **if** *(type=="CS")* **then**
22   n_coord = read(); i=i+1;
23   phi = angleWI(p_coord, c_coord, n_coord, boundary_set);
24   **if** *(phi<phi_min)* **then**
25   phi_min = phi; n_id = id;
26   **end**
27   **end**
28   **if** *(type=="SN")* **then**
29   **if** *(selected)* **then**
30   stop;
31   **else**
32   selected = true; boundary = true;
33   phi_min = 10; i=0; p_coord = read();
34   boundary_set = read();
35   send(c_id+"|"+"AC", *);
36   **end**
37   **end**
38 **until** *false*;

a fictitious node, itself and its neighbors. In our example, the next node is the light gray node (see Figure 4.19(b)). It will then send it a message to select it as the next boundary node, and also the value of its x-coordinate. Once the light gray node receives these messages, it will compare the received x-coordinate, which is the x-coordinate of the starting node, to its own x-coordinate. If it is smaller, which is the case in this example, then it will send a reset message to its previous node (see Figure 4.19(c)), here, to the dark gray node. After that, it will restart the D-LPCN algorithm (see Figures 4.19(d) and (e)). We will find ourselves in the same situation as previously, where the selected node has an x-coordinate smaller than the one of the starting node. In the same way, this node will send a reset message (see Figure 4.19(f)) and restart the D-LPCN algorithm (see Figure 4.19(g)). As we can see, three light gray nodes are visited. The two first visited nodes sent the x_{min} value, which is the x-coordinate of the starting node until visiting the third one having an x-coordinate smaller than x_{min}. This time, this node is the one having the global minimum x-coordinate (see Figure 4.19(h)), which means that there is no node having an x-coordinate smaller than its own x-coordinate. In this case, the D-LPCN will be executed normally (see Figure 4.19(i)) until coming back to the starting node.

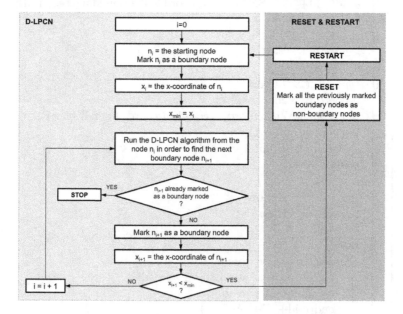

Figure 4.18: A flowchart of the D-RRLPCN algorithm.

Figure 4.18 shows the flowchart of the D-RRLPCN algorithm, and in order to present its algorithmic version, we define a message alphabet, described in Table 4.1, and the functions used in the algorithm, described in Table 4.2.

Finally, Algorithm 21 presents the main steps executed by each sensor node for the discovery of the boundary nodes in a WSN.

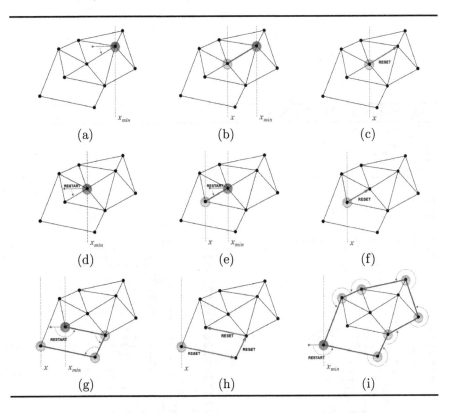

Figure 4.19: An illustration of the D-RRLPCN algorithm.

Algorithm 21: D-RRLPCN

```
 1  boundary = false;
 2  once = true;
 3  cid = getId();
 4  cx = getX();
 5  cy = getY();
 6  i = 0;
 7  n = getNumberOfNeighbors();
 8  nid = 0;
 9  nid_ref = -1;
10  phi_min = 10;
11  first = id of the starting node;
12  prevList = ∅;
13  repeat
14  |   if (cid==first and once) then
15  |   |   boundary = true;
16  |   |   once = false;
17  |   |   xmin = cx;
18  |   |   px = cx-1;
19  |   |   py = cy;
20  |   |   data = cid+"|"+"AC";
21  |   |   send(data, *);
22  |   end
23  |   id = read();
24  |   type = read();
25  |   if (type=="AC") then
26  |   |   data = cid+"|"+"CS"+"|"+cx+"|"+cy;
27  |   |   send(data, id);
28  |   end
29  |   if (type=="CS") then
30  |   |   id = read();
31  |   |   nx = read();
32  |   |   ny = read();
33  |   |   phi= angle(px,py,cx,cy,nx,ny);
34  |   end
35  |   if (phi<phi_min) then
36  |   |   phi_min = phi;
37  |   |   nid = id;
38  |   end
39  |   i = i + 1;
40  |   if (i==n) then
41  |   |   if (nid==nid_ref&cid==first) then
42  |   |   |   STOP;
43  |   |   else
44  |   |   |   if (nid_ref<0) then
45  |   |   |   |   nid_ref = nid;
46  |   |   |   end
47  |   |   end
48  |   |   i = 0;
49  |   |   data=cid+"|"+"SN"+"|"+cx+"|"+cy+"|"+xmin;

51  end
52  if (type=="SN") then
53  |   boundary = true;
54  |   n = getNumberOfNeighbors();
55  |   nid = 0;
56  |   phi_min = 10;
57  |   px = read();
58  |   py = read();
59  |   xmin = read();
60  |   if (cx<xmin) then
61  |   |   delay(t);
62  |   |   once = true;
63  |   |   first = cid;
64  |   |   data = cid+"|"+"DN"+"|"+first;
65  |   |   send(data, id);
66  |   else
67  |   |   prevList.add(id);
68  |   |   data = cid+"|"+"AC";
69  |   |   send(data, *);
70  |   end
71  end
72  if (type=="DN") then
73  |   boundary = false;
74  |   first = read();
75  |   if (prevList.size()>=0) then
76  |   |   data = cid+"|"+"DN"+"|"+first;
77  |   |   previous = prevList.getLast();
78  |   |   prevList.removeLast();
79  |   |   if (previous.hasPrevious()) then
80  |   |   |   send(data, previous);
81  |   |   end
82  |   end
83  end
50  until false;
84  send(data, nid);
```

Chapter 5

The simulator CupCarbon and boundary detection

There exist a number of tools that simulate distributed systems, but most of them are dedicated to a specific domain and, most importantly, they do not offer an easy-to-use graphical interface. Sometimes, users apply these tools to validate their newly conceived algorithms with respect to a given technological standard. However, their efforts are doubled, i.e., consist of the development of the algorithm itself and its application within the specific domain, also for the particular reason that these simulators can only be used like that. In this chapter, we present a new platform, called *CupCarbon*, which can be used to simulate and validate distributed algorithms in their theoretical version. It offers an interface allowing to visualize the result of an algorithm's execution. It is then possible to associate the algorithm to a given standard like the ZigBee protocol, used in the case of WSNs and IoT. CupCarbon has shown to be very useful for all aspects of algorithmic boundary detection.

This chapter is composed of 5 parts:

- **Part 1**: Presentation of the CupCarbon environment as composed of the graphical user interface and the map,

- **Part 2**: Presentation of the objects of CupCarbon, i.e., of those elements that can be added to the map of CupCarbon and how to manipulate and configure them,

- **Part 3**: Presentation of SenScript, the script used to program the nodes. Only those commands are presented that are used in the algorithms shown in this book,

- **Part 4**: Presentation of 12 scripts of algorithms that are frequently used and that represent the main parts of the D-LPCN algorithm,

- **Part 5**: Presentation of 4 versions of the D-LPCN algorithm, in which the starting node is determined automatically using one of the leader election algorithms presented in Chapter 4.

5.1   CupCarbon for network simulation

The simulation of networks is an essential tool for testing protocols and their performance prior to deployment. Researchers often use network simulators to test and validate proposed protocols and algorithms before their real deployment [117]. Indeed, such an establishment may be costly and challenging, especially for the distribution of a large number of nodes at a large scale [84, 104]. This is why the simulation of networks is essential, which justifies the number of existing simulators, like: TOSSIM [80], Avrora [111], OMNeT++ [113], SENSE [34] and NS-2 [65]. However, these simulators are not suitable for IoT networks and do not take into account the topology of the network, unlike the simulator that we will present in this chapter.

CupCarbon is a Smart City and Internet of Things Wireless Sensor Network (SCI-WSN) simulator [17, 18, 19, 20, 29, 87]. Its objective is to design, visualize, debug and validate distributed algorithms for monitoring, environmental data collection, etc., and to create environmental scenarios such as fires, gas, mobiles, and more generally, within educational and scientific projects. Not only can it help to visually explain the basic concepts of sensor networks and how they work; it may also support scientists to test their wireless topologies, protocols, etc.

CupCarbon offers two simulation environments. The first such environment enables the design of mobility scenarios and the generation of natural events such as fires or gas as well as the simulation of vehicles or flying objects such us UAVs, insects, etc. The second simulation environment represents a discrete event simulation of WSNs which takes into account the scenario designed on the basis of the first environment.

Networks can be designed and prototyped by an ergonomic and easy to use interface using the OpenStreetMap (OSM) framework[1] to deploy sensors directly on the map. It includes a script called SenScript, which allows to program and configure each sensor node individually. The model of the energy consumption of each radio module can be chosen from the integrated models and entered as an equation.

CupCarbon simulation is based on the application layer of the nodes. This makes it a real complement to existing simulators. It does not simulate all protocol layers due to the complex nature of urban networks which need to incorporate other complex and resource consuming information such as buildings, roads, mobility, signals, etc.

CupCarbon offers the possibility to simulate algorithms and scenarios in several steps. For example, there could be a step for determining the nodes of interest, followed by a step related to the nature of the communication between these nodes to perform a given task such as the detection of an

---

[1]http://www.openstreetmap.org

event, and finally, a step describing the nature of the routing to the base station in case that an event is detected. The current version of CupCarbon allows to configure the nodes dynamically in order to be able to split nodes into separate networks or to join different networks, a task which is based on the network addresses and the channel. The energy consumption can be calculated and displayed as a function of the simulated time. This allows to clarify the structure, feasibility and realistic implementation of a network before its real deployment. The propagation visibility and the interference models are integrated and include the ZigBee, LoRa and WiFi protocols.

5.2   The environment of CupCarbon

As shown by Figure 5.1, the CupCarbon Graphical User Interface (GUI) is composed of the following five main parts, which will be detailed below.

1. The map (in the center)

2. The menu bar (on the top)

3. The toolbar (under the menu)

4. The parameter menu (on the left)

5. The state bar (at the bottom)

6. The console (between the bottom and the center)

5.2.1   Menu bar

As shown by Figure 5.2, the menu bar is composed of 10 items:

1. Project: to create and open projects,

2. Edit: to execute edition operations like Copy/Paste, Undo/Redo, etc.,

3. Add: to add objects on the map like sensor nodes, mobiles, markers, etc.,

4. Display: to display or hide some elements on the map,

5. Selection: to select/deselect objects on the map,

6. Solver: to run integrated algorithms. These are mainly centralized algorithms applied to graphs,

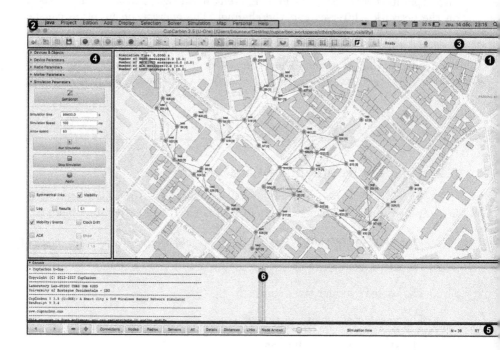

Figure 5.1: The CupCarbon Graphical User Interface.

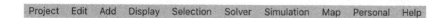

Figure 5.2: The menu bar of CupCarbon.

7. Simulation: to configure, run or stop a simulation, to open the script window, and to draw the energy consumption curve,

8. Map: to change the map,

9. Personal: to run user added algorithms,

10. Help: to access to help.

5.2.2 Map

The map is the main part of the graphical interface of CupCarbon. It is the part where the network and the objects of the project can be designed. The map can be changed according to the preference of the user or the way the information must be presented.

It allows to display the simulation time, the state of the simulation (running or stopping) as well as information about the number of exchanged messages in terms of number of sent, received, ACK (acknowledged) and lost messages or bits.

5.2.3 Toolbar

Figure 5.3: The toolbar of CupCarbon.

The toolbar (Figure 5.3) is used to quickly access the main actions on the map, like opening or saving a project, selecting/deselecting objects, etc.

5.2.4 Parameter menu

The parameter menu (Figure 5.4 (a)-(e)) contains 5 main panels:

1. Object list: used to easily select a single object or a list of objects according to their type,

2. Device parameters: used to configure selected devices on the map,

3. Radio parameters: used to configure, add or remove radio modules,

4. Marker parameters: used to add markers, routes or to create zones where random networks and buildings can be generated and loaded,

5. Simulation parameters: used to start, stop and configure the simulation.

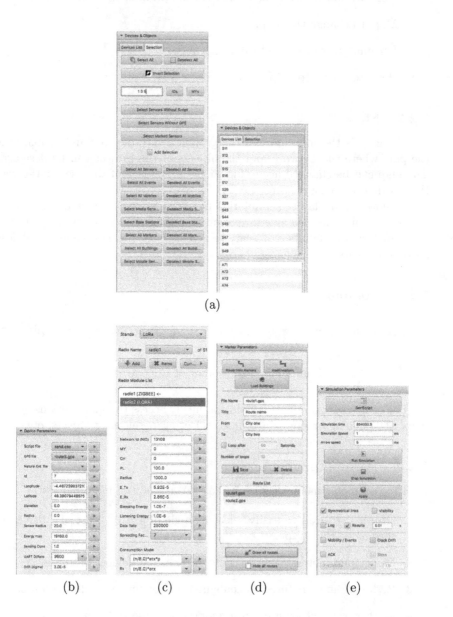

Figure 5.4: The parameter panels.

5.2.5   Console

The console is used by the simulator to display messages useful for the user during the simulation. It has two parts. The first is situated on the left and used to display messages related to the simulation. It is also possible to display messages of the nodes using the SenScript command `cprint`. The second part on the right is used to display errors that may arise during the simulation.

5.3   The objects of CupCarbon

This section presents the main objects that can be added on the map and that are necessary for the design and preparation of the simulation environment and scenario.

5.3.1   Sensor node

The sensor node is the main object of study of CupCarbon. As shown by Figure 5.5, it contains three main parts: radio modules, a sensing unit and a battery. In the center of the sensor node, we find the name S followed by its id. For example, if its identifier id is equal to 4, then its name will be S4. If a SenScript is assigned to it, then it will be displayed in gray color above its name. The print messages will be displayed below its name.

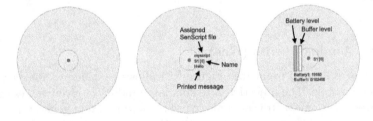

Figure 5.5: Sensor node parts.

A sensor node can contain many radio modules with different or identical standards (802.15.4, WiFi or LoRa). However, for communication, only one radio module is considered. The great circle represents the radio range of the current radio (see Figure 5.5). The current radio can be changed during the simulation by using the command **radio** in the SenScript file (cf. SenScript user reference [20]). A sensor node contains a sensing unit represented by a transparent white circle (see Figure 5.6).

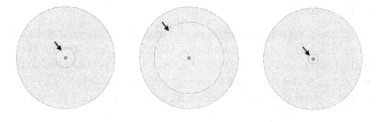

Figure 5.6: Sensing area.

5.3.2   Base station (Sink)

The base station is identical to a sensor node with the exception that it has an infinite battery (see Figure 5.7).

Figure 5.7: Base station.

5.3.3   Analog events (Gas)

Gas or a natural event (see Figure 5.8) allows to generate analog values. It can be mobile. The objective is to simulate random or given values coming from the environment. It is possible to use the natural event generator in order to generate random values based on the Gaussian distribution. To take into account the event during the simulation process, the (Mobility/Event) box must be activated in the "Simulation Parameters" view.

Figure 5.8: Analog event (gas).

5.3.4 Mobile

A mobile (see Figure 5.9) is used to generate digital events (detection). It depends on its route (trajectory) which can be created by markers. During the simulation, the mobile moves each second to the next location on the route, and can then go back to the initial point.

Figure 5.9: A mobile.

5.3.5 Marker

The markers can be used for different objectives, such as:

- Adding sensor nodes randomly (see Figure 5.10),

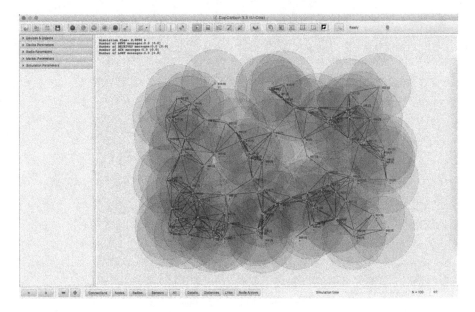

Figure 5.10: A random network.

- Adding sensor nodes (see Figure 5.11),
- Generating routes (see Figure 5.12),
- Adding buildings (see Figure 5.13),

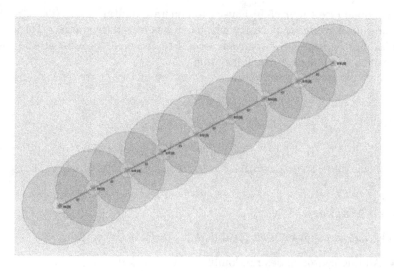

Figure 5.11: Markers are transformed into sensor nodes.

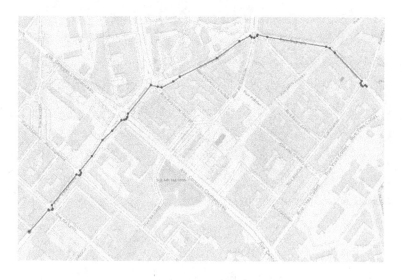

Figure 5.12: Generating a route.

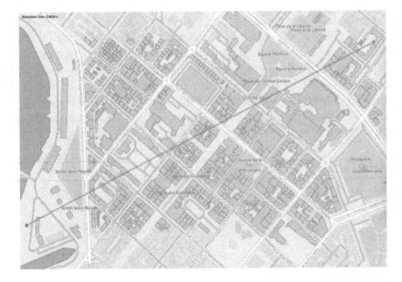

Figure 5.13: Adding buildings.

• Drawing buildings or obstacles (see Figure 5.14).

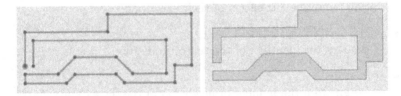

Figure 5.14: Drawing a building or an obstacle.

5.4 An introduction to SenScript

SenScript is the script used to program sensor nodes of the CupCarbon simulator. It is a script in which variables are not declared, but can be initialized. For string variables, it is not necessary to use the quotes. A variable is used by its name (e.g., x), and its value is determined by $ (e.g., $x). Algorithm 22 shows an example of a SenScript code. The command `atget id` of line 1 allows to assign to the variable `cid` the identifier of the current sensor node. The command `loop` allows to start the `loop`, whose body will be executed infinitely. The command `wait` allows to wait for reception of a message in the buffer. This command is blocking and the next code will not be executed until at least one message is received. The command `read` of line 4 will assign the received message to x. In line 5, we test if the received value (an identifier) is smaller than the value of the current identifier `cid`. If this is the case, line 6 will be executed, and the sensor node will be marked (`mark 1`), otherwise (line 7), line 8 will be executed, and the sensor node will be unmarked (`mark 0`).

Algorithm 22: SenScript example

```
1  atget id cid;
2  loop;
3     wait;
4     read x;
5     if($x < $cid)
6        mark 1;
7     else
8        mark 0;
9  end
```

In the following, we will introduce a number of important commands of SenScript and describe the effects of their execution:

- `atid v`
  → changes the identifier of the sensor node to v.

- `atnd n`
  → assigns to n the number of neighbors of the sensor node.

- `atnd n v`
  → assigns to n the number of neighbors of the sensor node and to v the vector of the neighbors' identifiers.

- `atget id v`
  → assigns to v the identifier of the sensor node.

- `data d v1 v2 ...`
  → assigns to d the string "v1#v2#...". The command `data` is used to create messages to send.

- `delay v`
  → creates a delay of v milliseconds before going to the next instruction.

- `edge a v`
  → marks (a=1) or unmarks (a=0) the edge (communication link) between the current sensor node and the node having the identifier v.

- `mark v`
  $\rightarrow$ marks (v=1) or unmarks (v=0) the current sensor node.

- `rdata v a b ...`
  $\rightarrow$ we assume that p is a message formed by using the command `data`. It is possible to form p manually as a string containing a `#` separator (example: p=hello#4). In this example, `rdata $p a b` will lead to a=hello and b=4.

- `read x`
  $\rightarrow$ assigns the value in the buffer (the received message) to x.

- `script fileName`
  $\rightarrow$ loads the script file with the name fileName.

- `send`
  $\rightarrow$ used to send messages.
  `send hello 2`
  $\rightarrow$ sends the message `hello` to the sensor node having id=2 in a unicast mode.
  `send hello` or `send hello`
  $\rightarrow$ sends the message `hello` in a broadcast mode.

- `wait [t]`
  `wait`
  $\rightarrow$ creates a delay until receiving data in the buffer
  `wait 1000`
  $\rightarrow$ creates a delay until receiving data in the buffer. If no data is received after 1 second (=1000 milliseconds), the execution of the script will be continued.

- `stop`
  $\rightarrow$ stops the execution.

5.5 SenScript examples

In this section, we will present some basic and important algorithms necessary to formulate the D-LPCN algorithm, used to determine the boundary nodes of distributed systems.

5.5.1 Sending and receiving messages

The example of sending and receiving messages is very important since the main actions of the simulation under CupCarbon are based on sending/receiving messages. As an example, let us consider a sensor node that will send infinite messages A and B with a delay of one second to another sensor node. The receiver will be marked if it receives A, and unmarked if it receives B.

```
// Transmitter
1: loop
2: send A 2
3: delay 1000
4: send B 2
5: delay 1000

// Receiver
1: loop
2: wait
3: read v
4: if($v==A)
5:    mark 1
6: else
7:    mark 0
8: end
```

5.5.2 Routing

Here we consider 4 sensor nodes amongst which messages sent by sensor node S1 will be routed to sensor node S4 (see Figure 5.15).

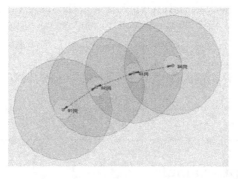

Figure 5.15: Routing messages from S1 to S4.

In the same way as in the previous example, we can create codes for the transmitter (sensor node S1) and the receiver (sensor node S4). For this, we will add two routers. The first one, sensor node S2, will route the received messages to sensor node S3, and the second router S3 will route the received messages to sensor node S4 (the receiver). The codes of the routers are given as follows:

```
// Router 1 (S2)
1: loop
2: wait
3: read v
4: send $v 3

// Router 2 (S3)
1: loop
2: wait
3: read v
4: send $v 4
```

5.5.3 Flooding

The Flooding algorithm represents a distributed version of Breadth-First Search (BFS) allowing to find a tree in a graph [101]. It starts from a given node (the root) which will send, in a broadcast, a message with its identifier. Each of the neighbors receiving that message will save locally the identifier of the root and send itself, in a broadcast, a message with its own identifier. No further received message will be taken into account. The other nodes that receive a message, will save the identifier of the first transmitter once and send themselves in a broadcast a message with their identifier. This process stops as soon as each node has received at least one message. The following code shows the SenScript code of the flooding process. The root node, the one having the identifier id_root, will send a message with its identifier id (lines 3 to 6). The other nodes will wait for the reception of a rid message (lines 8 and 9). Once a message is received, they fix their previous node (their parent) by assigning its value to the variable pid (line 11) and then send their own identifier id in a broadcast (line 12). This process is done once (line 14).

```
// Flooding
01: atget id id
02: set once 0
03: if($id==id_root)
04:    send $id
05:    stop
06: end
07: loop
08: wait
09: read rid
10: if($once==0)
11:    set pid $rid
12:    send $id
13:    set once 1
14: end
```

5.5.4   Flooding for Leaf Finding (FLF)

The Flooding for Leaf Finding (FLF) algorithm is based on the previously presented flooding algorithm. Its main objective is to find the leaves of the tree obtained by flooding. To do this and before sending a broadcast message, a node will send an acknowledgment to its parent node. Then each node, after sending a broadcast message, will wait a given time for an acknowledgment. If there is no acknowledgment received, the node will be considered as a leaf.

The following algorithm shows the SenScript code of the FLF algorithm. It is similar to the one given above. At the beginning, each node will be considered as a leaf (line 3). In this algorithm the nodes can send two types of messages. The first is the broadcast message A (lines 14 and 15) and the second is the acknowledgment message B (lines 16 and 17). If a node receives a message B it will be considered as a non-leaf (lines 21 to 23).

```
// FLF
01 : atget id id
02 : set once 0
03 : mark 1
04 : if($id==id_root)
05 :     data message A $id
06 :     send $m
07 : end
08 : loop
09 : wait
10 : read m
11 : rdata $message t rid
12 : if($t==A)
13 :     if($once==0)
14 :         data message A $id
15 :         send $message
16 :         data message B $id
17 :         send $message $rid
18 :         set once 1
19 :     end
20 : end
21 : if($t==B)
22 :     mark 0
23 : end
```

5.5.5   Wait-Before-Starting (WBS)

The Wait-Before-Starting (WBS) algorithm allows to select the node having the minimum value [22]. This value can be deterministic or generated randomly, and it can be any parameter of a sensor node such as the identifier, the current level of the battery, the x- or y-coordinate, the signal power, etc. Its concept is based on waiting for a time which is proportional to the value. The node having the minimum value will be the one which will wait the least. The waiting time should be long enough to allow all the other nodes to receive the message before the end of their waiting times. The SenScript program of this algorithm is given by the following code, which determines the node having the minimum identifier. In line 3, the waiting time w is determined, and in this example, it is proportional to 1 second (1000 milliseconds). This means that the node with identifier 1 will wait for 1 second. Each node starts waiting (line 4). The node with the minimum identifier will be the first to finish waiting and not to receive any further messages (lines 5 and 6). In this case, it will be marked as the one having the minimum identifier (line 7) and start the flooding process in order to inform all the other nodes (lines 8), which in turn are used to run the FLF algorithm.

```
// WBS
01 : atget id id
02 : set once 0
03 : set w $id*1000
04 : wait $w
05 : read m
06 : if($m==\)
07 :     mark 1
08 :     send A
09 :     set once 1
10 : end
11 : loop
12 : if($once==0)
13 :     send A
14 :     set once 1
15 : end
16 : stop
```

5.5.6   Wait-Before-Starting with Flooding

The following SenScript algorithm is just a combination of the Flooding and the WBS algorithms, where the starting node (root) in the flooding algorithm is determined using the WBS algorithm.

```
// WBS-Flooding
01 : atget id id
02 : set once 0
03 : set w $id*1000
04 : wait $w
05 : read rid
```

```
06 : if($rid==\)
07 :     send $id
08 :     stop
09 : end
10 : loop
11 : if($once==0)
12 :     set once 1
13 :     send $id
14 : end
15 : stop
```

5.5.7   Wait-Before-Starting with FLF

The following SenScript algorithm is just a combination of the WBS and the FLF algorithms, where the starting node (root) in the FLF algorithm is determined using the WBS algorithm.

```
// WBS-FLF
01 : atget id id
02 : set once 0
03 : mark 1
04 : set w $id*1000
05 : wait $w
06 : read m
07 : if($m==\)
08 :     data m A $id
09 :     send $m
10 : end
11 : loop
12 : rdata $m t rid
13 : if($t==A)
14 :     if($once==0)
15 :         data m A $id
16 :         send $m
17 :         data m B $id
18 :         send $m $rid
19 :         set once 1
20 :     end
21 : end
22 : if($t==B)
23 :     mark 0
24 : end
25 : stop
```

5.5.8   Local Minima Finding

The following algorithm allows to find the local minima in a network. We assume that each node will generate a random value between 0 and 1000. The algorithm will find out whether a node is a local minimum, which means the one with minimum value among its neighbors. At the beginning, each node will be considered as a local minimum (line 2) and will send in a broadcast mode its value to its neighbors (line 3). Then it will compare this value with the values received from its neighbors. If such a value is smaller than its own (line 7), this node will be considered as a non-local minimum (line 8).

```
// Local Min
01: randb xmin 0 1000
02: set leader 1
03: send $xmin
04: loop
05: wait
06: read x
07: if ($x < $xmin)
08:     set leader 0
09: end
```

5.5.9   Global Minimum Finding

The following algorithm allows to find the global minimum in a network. It is based on the Local Minima Finding algorithm, presented above, where instead of sending only one broadcast message, a node sends a broadcast message each time it receives a value which is smaller than its own (line 10). This last one will be updated by each smaller received one (line 8). This process will stop after a given time. The node having the smallest value will be the one which will not have received any smaller value.

```
// Global Min
01: randb xmin 0 1000
02: set leader 1
03: send $xmin
04: loop
05: wait
06: read x
07: if ($x < $xmin)
08:     set leader 0
09:     set xmin $x
10:     send $xmin
11: end
```

5.5.10   The R-LOGO algorithm

In the following algorithms, we use different types of messages (T1 to T4). These types are detailed in Section 4.4. The Revised Local Optima to Global Optimum (R-LOGO) algorithm starts with the WBS algorithm (Section 5.5.5) to determine the root node (line 8 of the R-LOGO1 script) and the non-root nodes (line 10 of the R-LOGO1 script). Then the root node will execute the script R-LOGO2 and the non-root nodes will execute the script R-LOGO3. In this example, we want to find the leader given by the node with minimum x-coordinate. At the beginning, each node will be considered as the leader (line 2 of R-LOGO2 and line 2 of R-LOGO3). Then, if there is no received message after 1 second (line 5 of R-LOGO3), which means that the root and the non-root nodes have been determined, each node except the root starts the Local Minima Finding algorithm (Section 5.5.8) by sending broadcast messages of type T1 (lines 6 and 7 of R-LOGO3) and receiving messages of the same type (lines 10 to 17 of R-LOGO3). After another given time, fixed to 3 seconds (line 3 of R-LOGO2), if there is no message received, the local minima are determined. The root node will start the flooding process by sending a message of type T2 asking each local minimum to send it its identifier together with its x-coordinate (lines 4 and 5 of R-LOGO2 and lines 18 to 29 of R-LOGO3). Then each local minimum will send over the path or the branch determined by the flooding process a message of type T3 to the root (lines 30 to 35 of R-LOGO3). During this process a stack is built in order to send the path formed by nodes situated between each local minimum and the root (line 32 of R-LOGO3). Each time a T3 message is received by the root, the received value will be compared in order to determine which local minimum is the global one (lines 24 to 32 of R-LOGO2). If no message is received after 5 seconds (lines 9 and 14 of R-LOGO2), all the values of the local minima have been received and compared, in other words, the global minimum has been determined, the root will send a message of type T4 to the global minimum in order to elect it (lines 14 to 22 of R-LOGO2 and lines 36 to 46 of R-LOGO3). The condition of line 15 of R-LOGO2 is used to determine whether there is or not a local minimum smaller than the value of the root. In case that there is such a value, the root will be the leader.

```
// R-LOGO1 (starting algorithm)
01 : atget id id
02 : set w $id*1000
03 : loop
04 : wait $w
05 : read message
06 : send A
07 : if($message==\)
08 :     script root_logo
09 : else
```

```
10 :     script non_root_logo
11 : end

// R-LOGO2 (root node)
01 : getpos2 xmin y
02 : set leader 1
03 : delay 3000
04 : data message T2 $id
05 : send $message
06 : set twait 0
07 : loop
08 : if($twait==0)
09 :     wait 5000
10 : else
11 :     wait
12 : end
13 : read message
14 : if($message==\)
15 :     if($leader==0)
16 :         spop nid tmin
17 :         data message T4 $idmin $tmin
18 :         send $message $nid
19 :         set twait 1
20 :     else
21 :         stop
22 :     end
23 : else
24 :     rdata $message type rid rx t
25 :     if(($type==T3) && ($rx<$xmin))
26 :         set leader 0
27 :         mark 0
28 :         set idmin $rid
29 :         set xmin $rx
30 :         set tmin $t
31 :     end
32 : end

// R-LOGO3 (non-root node)
01 : getpos2 xmin y
02 : set leader 1
03 : set once1 0
04 : set once2 0
05 : delay 1000
06 : data message T1 $id $xmin
07 : send $message
```

```
08 : loop
09 : wait
10 : read message
11 : rdata $message type rid v
12 : if($type==T1)
13 :     if($v<$xmin)
14 :         set xmin $v
15 :         set leader 0
16 :     end
17 : end
18 : if(($type==T2) && ($once1==0))
19 :     set once1 1
20 :     set prev $rid
21 :     data message T2 $id
22 :     send $message
23 :     if($leader==1)
24 :         set leader 0
25 :         data message T3 $id $xmin $id
26 :         send $message $prev
27 :         delay 1000
28 :     end
29 : end
30 : if($type==T3)
31 :     rdata $message type rid v t
32 :     sadd $id t
33 :     data message T3 $rid $v $t
34 :     send $message $prev
35 : end
36 : if(($type==T4) && ($once2==0))
37 :     set once2 1
38 :     rdata $message type idmin t
39 :     if($idmin==$id)
40 :         set leader 1
41 :     else
42 :         spop nid t
43 :         data message T4 $rid $t
44 :         send $message $nid
45 :     end
46 : end
```

5.5.11 The R-BrOGO algorithm

The Revised Branch Optima to Global Optimum (R-BrOGO) algorithm starts with the WBS algorithm (Section 5.5.5) to determine the root node (line 7 of R-BrOGO1) and the non-root nodes (line 9 of R-BrOGO1). The root node will inform the other nodes using the flooding process. Since the first step of the BrOGO algorithm is flooding, too, it will be used once for the WBS step and also to start the BrOGO algorithm. Then the root node will execute the script R-BrOGO2 and the non-root nodes will execute R-BrOGO3. In this example, we want to find the leader given by the minimum x-coordinate. At the beginning, only the root node will be considered as the leader (line 4 of R-BrOGO2 and line 8 of R-BrOGO3). The BrOGO algorithm starts with the FLF algorithm as presented in Section 5.5.4 using messages of types T1 and T2 (lines 5 and 6 of R-BrOGO2 and lines 18 to 30 of R-BrOGO3). If none of the leaves has received a message after 1 second, they will send a message of type T3 to the root (lines 10 to 16 of R-BrOGO3) in order to send the minimum value, its branch and its corresponding node identifier (lines 31 to 40 of R-BrOGO3). The root will then compare the minima received from all branches (lines 20 to 27 of R-BrOGO2) and determine the node and the branch containing the global minimum. If after 3 seconds (line 10 of R-BrOGO2) the root has not received any value, it will send a message of type T4 to elect the leader (lines 15 to 19 of R-BrOGO2 and lines 41 to 50 of R-BrOGO3).

```
// R-BrOGO1 (starting algorithm)
01 : atget id id
02 : set w $id*1000
03 : loop
04 : wait $w
05 : read message
06 : if($message==\)
07 :     script root_brogo
08 : else
09 :     script non_root_brogo
10 : end

// R-BrOGO2 (root node)
01 : getpos2 xmin y
02 : set idmin $id
03 : set twait 0
04 : set leader 1
05 : data message T1 $id
06 : send $message
07 : loop
08 : mark $leader
09 : if($twait==0)
10 :     wait 3000
11 : else
12 :     wait
13 : end
```

```
14 : read message
15 : if($message==\)
16 :    data message T4
17 :    send $message $idmin
18 :    set twait 1
19 : end
20 : rdata $message type rid rx v
21 : if($type==T3)
22 :    if($rx<$xmin)
23 :       set leader 0
24 :       set xmin $rx
25 :       set idmin $rid
26 :    end
27 : end

// R-Br0G03 (non-root node)
01 : getpos2 x y
02 : set once1 0
03 : set once2 0
04 : set twait 0
05 : set brmin 1
06 : set idmin 0
07 : set leaf 1
08 : set leader 0
09 : loop
10 : if($message==\)
11 :    if($leaf==1)
12 :       set brmin 1
13 :       set twait 1
14 :       data message T3 $id $xmin 1
15 :       send $message $prev
16 :    end
17 : else
18 :    rdata $message type rid rx v
19 :    if(($type==T1) && ($once1==0))
20 :       set prev $rid
21 :       set once1 1
22 :       data message T2 $id
23 :       send $message $rid
24 :       data message T1 $id
25 :       send $message
26 :    end
27 :    if($type==T2)
28 :       set nid $rid
29 :       set leaf 0
```

```
30 :     end
31 :     if($type==T3)
32 :         if($rx<$xmin)
33 :             set xmin $rx
34 :             set idmin $rid
35 :             set brmin 0
36 :         end
37 :         set twait 1
38 :         data message T3 $id $xmin $v
39 :         send $message $prev
40 :     end
41 :     if($type==T4)
42 :         set twait 1
43 :         if($brmin==1)
44 :             set leader 1
45 :         else
46 :             data message T4
47 :             send $message $idmin
48 :         end
49 :     end
50 : end
51 : mark $leader
52 : if($twait==0)
53 :     wait 1000
54 : else
55 :     wait
56 : end
57 : read message
```

5.5.12   The DoTRo algorithm

The Dominating Tree Routing (DoTRo) algorithm starts with the Local
Minima Finding algorithm presented in Section 5.5.8 to determine each local
minimum [26]. In the following DoTRo code there are two steps. The first step
consists in determining the local minima (lines 5 and 6 and lines 18 to 22). If
after 1 second a node has not received any message (line 11) it goes to step 2
(line 12), and if it is a local minimum, it starts the flooding process (lines 13
to 16) by sending its value. In step 2 (line 24) and during the flooding process
started from each local minimum, each node will continue the flooding if it
receives a value which is smaller than its own (lines 33 to 40), otherwise, it
will do nothing. However, if after 3 seconds (line 25) there is no message received, the flooding process is done, and the leader, i.e., the global minimum,
is determined and will be marked (lines 27 to 31).

```
// DoTRo
01 : atget id cid
02 : getpos2 xmin y
03 : set step 1
04 : set leader 1
05 : data message $cid $xmin
06 : send $message
07 : loop
08 : if($step==1)
09 :     wait 1000
10 :     read message
11 :     if($message =\)
12 :         set step 2
13 :         if($leader==1)
14 :             data message $cid $xmin
15 :             send $message
16 :         end
17 :     else
18 :         rdata $message rid rx
19 :         if($rx<$xmin)
20 :             set leader 0
21 :         end
22 :     end
23 : end
24 : if($step==2)
25 :     wait 3000
26 :     read message
27 :     if($message ==\)
28 :         if($leader==1)
29 :             mark 1
30 :         end
31 :         stop
32 :     else
33 :         rdata $message rid rx
34 :         if($rx<$xmin)
35 :             set leader 0
36 :             set xmin $rx
37 :             data message $cid $rx
38 :             send $message
39 :         end
40 :     end
41 : end
```

5.6 SenScript of the D-LPCN algorithm

In this section, we will present the script of the algorithm D-LPCN. We first present the classical version where the starting node is determined manually. Then we will describe three other versions in which the starting node is determined using the following previously presented leader election algorithms: Minimum Finding, DoTRo and BrOGO. Note that, for simplicity, we will not consider the special situations presented in the original version of the D-LPCN algorithm (cf. Chapter 4).

5.6.1 Version 1: fixing the starting node manually

The D-LPCN algorithm allows to find the boundary nodes of a network. It starts from any boundary sensor node (for example, the one on the extreme left). Then each node calculates the angles formed by the previously found boundary node and each of its neighbors, and chooses the node that forms the smallest angle as the next boundary node. To write the script of this algorithm, we will use 3 types of messages. A message AC to ask the neighbors to send their coordinates, a message CS to send the coordinates and a message SN to inform a sensor that it is a boundary node. The script of the D-LPCN algorithm is given below. First, we fix the value of the starting node (line 1). As an example, we consider the network depicted in Figure 5.16 in which we assign the value 49 to the variable `first`, which means that the node with the identifier 49 will start the D-LPCN algorithm. We assume that this node is situated on the extreme left as shown by Figure 5.15. Then each sensor node will assign its current identifier (line 2) to the variable `cid` and calculate its current GPS coordinates `cx` and `cy` (line 3). The variable `nid` representing the boundary node following the current one is fixed to -1 (line 4) because at the beginning, each node does not know its subsequent boundary node. Lines 6 to 11 are executed only by the starting node (`cid=49`) and just once. That is why, in line 7, the variable `first` is set to -1. To calculate the angle formed by the fictitious boundary node and its neighbors (lines 8 to 11), we assume that there is a virtual sensor node, situated to the left part of the starting node, and having the coordinates (`px=cx-1` and `py=cy`). Then a message of type SN is specified (line 11) to go directly to line 40. The starting node will be marked as a boundary node (line 42) and start the process of finding the minimum angle (line 43) formed with its neighbors. This process begins with an angle of 10 radian. An `ATND` message will be sent to determine the number of the neighbors, required to know the number of CS messages to receive after sending the AC command (lines 46 and 47) which is used to ask the neighbors to send their coordinates. For each CS message received from a neighbor (line 21) and containing its coordinates (line 22), the starting node will calculate the angle formed with it (line 23) and save the angle and the identifier of the node forming the smallest angle in the vector **m** (lines 24 to 26). As soon as all the neighbors have sent their coordinates (line 28) the starting node will send a message SN to the neighbor forming the smallest angle (lines 30, 36 and 37) to select it as the next boundary node. The node which receives this SN message will do exactly the same process as the starting node in order to determine the next boundary node, and so on. Note that if the starting node selects a second time the same next boundary node (line 32), then the process will stop. This stop condition is explained in detail in Section 1.11.3.2.

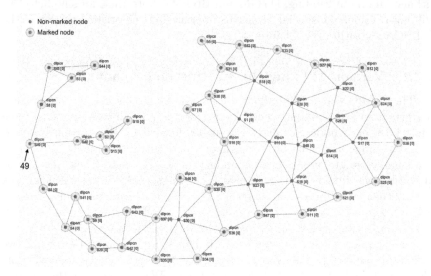

Figure 5.16: An illustration of the D-LPCN algorithm.

```
// D-LPCN
01 : set first 49
02 : atget id cid
03 : getpos2 cx cy
04 : set nid -1
05 : loop
06 : if ($cid==$first)
07 :     set first -1
08 :     set px $cx-1
09 :     set py $cy
10 :     data p 0 $type $px $py
11 :     set type SN
12 : else
13 :     wait
14 :     read p
15 :     rdata $p id type
16 : end
```

```
17 : if ($type==AC)
18 :     data p $cid CS $cx $cy
19 :     send $p $id
20 : end
21 : if ($type==CS)
22 :     rdata $p id type x y
23 :     angle2 a $px $py $cx $cy $x $y
24 :     data p $id $a
25 :     smin m $m $p
26 :     rdata $m id a
27 :     inc i
28 :     if($i==$n)
29 :        if($nid==-1)
30 :           set nid $id
31 :        else
32 :           if($nid==$id)
33 :              stop
34 :           end
35 :        end
36 :        data p $cid SN $cx $cy
37 :        send $p $id
38 :     end
39 : end
40 : if ($type==SN)
41 :     rdata $p id type px py
42 :     mark 1
43 :     data m $cid 10
44 :     atnd n
45 :     set i 0
46 :     data p $cid AC
47 :     send $p
48 : end
```

5.6.2 Version 2: starting from Minimum Finding

In this version, we will modify the Minimum Finding algorithm presented in Section 5.5.9 and use the command script to call the D-LPCN algorithm as soon as the extreme left node is found. In this case, it is not necessary to fix the value of the starting node, and line 1 of the D-LPCN algorithm given above must be removed. This script should be saved under the name dlpcn.csc in order to be able to call it from the Minimum Finding script using the command script dlpcn. Then in the Minimum Finding algorithm, we will call the previous script of D-LPCN, once finished, using the command script (line 16). This command is called by a node if it does not receive any message after 2 seconds, which means that the process of finding the starting node is completed. Note that we will add an additional delay of 1 second as soon as the starting node has been found (line 12) to allow the other nodes to finish their Minimum Finding algorithm and to be ready to start the D-LPCN algorithm. The new version of the Minimum Finding algorithm calling D-LPCN is given as follows:

```
// D-LPCN and Minimum Finding
01 : atget id cid
02 : getpos2 vmin y
03 : set leader 1
04 : send $vmin
05 : loop
06 : mark $leader
07 : wait 2000
08 : read v
09 : if($v == \)
10 :     if ($leader == 1)
11 :         set first $cid
12 :         delay 1000
13 :     else
14 :         set first -3
15 :     end
16 :     script dlpcn
17 : else
18 :     if ($v < $vmin)
19 :         set leader 0
20 :         set vmin $v
21 :         send $v
22 :     end
23 : end
```

5.6.3   Version 3: starting from R-BrOGO

To start the D-LPCN algorithm from a node found by the R-BrOGO algorithm presented above, we need to fix the value of the variable first according to the nature of the node. If it is a leader, then first=id (line 22 of R-BrOGO2 and line 56 of R-BrOGO3), where id is the identifier of the current node. If it is not a leader, then first=-1 (line 20 of R-BrOGO2 and line 59 of R-BrOGO3). Once the leader process is finished, the root node waits for 1 second (line 24 of R-BrOGO2) and sends a broadcast message of type T5 (defined in Section 4.4) to inform all the nodes that it is the leader (lines 25 and 26 of R-BrOGO2) using the flooding algorithm. After waiting for 2 seconds (lines 27 and 28 of R-BrOGO2) it starts the D-LPCN algorithm. During the flooding process, each node will receive a T5 message (line 51 of R-BrOGO3) and fix the value of its variable first according to its nature, i.e., `first=id` if it is a leader (line 56 of R-BrOGO3) and `first=-1`, otherwise (line 59 of R-BrOGO3). If the non-root node is a leader, it will wait for 2 seconds (line 57 of R-BrOGO3) and start the D-LPCN algorithm (line 61 of R-BrOGO3).

```
// R-BrOGO1 (starting algorithm)
01 : atget id id
02 : set w $id*1000
03 : loop
04 : wait $w
05 : read message
06 : if($message==\)
07 :     script root_brogo
08 : else
09 :     script non_root_brogo
10 : end
```

```
// R-BrOGO2 (root node)
01 : getpos2 xmin y
02 : set idmin $id
03 : set twait 0
04 : set leader 1
05 : data message T1 $id
06 : send $message
07 : loop
08 : mark $leader
09 : if($twait==0)
10 :    wait 3000
11 : else
12 :    wait
13 : end
14 : read message
15 : if($message==\)
16 :     set twait 1
17 :     if($leader==0)
18 :         data message T4
19 :         send $message $idmin
20 :         set first -1
21 :     else
22 :         set first $id
23 :     end
24 :     delay 1000
25 :     data message T5
26 :     send $message
27 :     delay 2000
```

```
28 :     script dlpcn
29 : end
30 : rdata $message type rid rx v
31 : if($type==T3)
32 :     if($rx<$xmin)
33 :         set leader 0
34 :         set xmin $rx
35 :         set idmin $rid
36 :     end
37 : end


// R-BrOGO3 (non-root node)
01 : getpos2 xmin y
02 : set once1 0
03 : set once2 0
04 : set twait 0
05 : set brmin 1
06 : set idmin 0
07 : set leaf 1
08 : set leader 0
09 : loop
10 : mark $leader
11 : if($message==\)
12 :     if($leaf==1)
13 :         set brmin 1
14 :         set twait 1
15 :         data message T3 $id $xmin 1
16 :         send $message $prev
17 :     end
18 : else
19 :     rdata $message type rid rx v
20 :     if(($type==T1) && ($once1==0))
21 :         set prev $rid
22 :         set once1 1
23 :         data message T2 $id
24 :         send $message $rid
25 :         data message T1 $id
26 :         send $message
27 :     end
28 :     if($type==T2)
29 :         set nid $rid
30 :         set leaf 0
31 :     end
32 :     if($type==T3)
33 :         if($rx<$xmin)
```

```
34 :              set xmin $rx
35 :              set idmin $rid
36 :              set brmin 0
37 :          end
38 :          set twait 1
39 :          data message T3 $id $xmin $v
40 :          send $message $prev
41 :      end
42 :      if($type==T4)
43 :          set twait 1
44 :          if($brmin==1)
45 :              set leader 1
46 :          else
47 :              data message T4
48 :              send $message $idmin
49 :          end
50 :      end
51 :      if(($type==T5) && ($once2==0))
52 :          set once2 1
53 :          data message T5
54 :          send $message
55 :          if($leader==1)
56 :              set first $id
57 :              delay 2000
58 :          else
59 :              set first -1
60 :          end
61 :          script dlpcn
62 :      end
63 : end
64 : if($twait==0)
65 :      wait 1000
66 : else
67 :      wait
68 : end
69 : read message
```

5.6.4   Version 4: starting from DoTRo

DoTRo is the best algorithm to be combined with D-LPCN because it is easy to implement and very few changes in the original D-LPCN algorithm are needed. Once the leader (the node with minimum x-coordinate) is determined, all the other nodes are in the situation of waiting for messages and ready to start a new algorithm. Then, in the DoTRo algorithm, we just need to call the D-LPCN algorithm as soon as the leader has been determined (line

37 of the following script) and the value of the variable **first** has been fixed (lines 31 and 32) either to the value of cid, the identifier of the leader, or to -1 for the non-leader nodes (line 7).

```
// D-LPCN and DoTRo
01 : atget id cid
02 : getpos2 xmin y
03 : set step 1
04 : set leader 1
05 : data message $cid $xmin
06 : send $message
07 : set first -1
08 : loop
09 : if($step==1)
10 :     wait 1000
11 :     read message
12 :     if($message==\)
13 :         set step 2
14 :         if($leader==1)
15 :             data message $cid $xmin
16 :             send $message
17 :         end
18 :     else
19 :         rdata $message rid rx
20 :         if($rx<$xmin)
21 :             set leader 0
22 :         end
23 :     end
24 : end
25 : if($step==2)
26 :     wait 3000
27 :     read message
28 :     if($message==\)
29 :         if($leader==1)
30 :             mark 1
31 :             set first $cid
32 :             delay 1000
33 :         end
34 :         script dlpcn
35 :     else
36 :         rdata $message rid rx
37 :         if($rx<$xmin)
38 :             set leader 0
39 :             set xmin $rx
40 :             data message $cid $rx
```

```
41 :            send $message
42 :        end
43 :    end
44 : end
```

Chapter 6

Applications

In this chapter, we will present some applications for which some of the presented algorithms can be used, especially LPCN and D-LPCN. We will also show how to adapt LPCN or RR-LPCN to the case of pixel graphs, where instead of choosing pixels based on their polar angles, we can just choose them based on their polar order. Our first example shows how to find boundary nodes of a Wireless Sensor Network (WSN) and how to detect whether a node on the boundary is faulty. The second example explains how to find gaps and voids in WSNs. The third one is on the extraction of complex clusters in a set of two-dimensional data, and the last one shows how to draw a contour of a zone of interest within an image, application that we highlight by an example from medical image processing and one from fingerprint detection. Finally, we conclude the chapter by presenting the concept of angle graphs.

6.1 Finding the boundary nodes of a WSN

Determining the boundary of a WSN is at the heart of our approach and has strongly motivated the writing of this book. Before presenting the algorithmic details, let us review some elementary concepts of WSNs. Figure 6.1 shows a wireless sensor node composed of the following main elements:

- Microcontroller (for example, an Arduino card)

- Battery (lithium polymer)

- Radio module (for example, a ZigBee based protocol)

- Sensor unit (for example, a motion sensor)

It is possible to integrate many radio modules and many sensor units into such a sensor node. For simplicity and throughout this chapter, we will assume that a sensor node has only one radio module and one sensor unit.

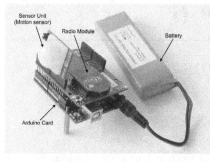

Figure 6.1: Main components of a wireless sensor node.

As illustrated schematically in Figure 6.2, a sensor node can be represented as the center of two circular zones. The white zone of a sensor node represents its radio range. A sensor node can communicate with any other sensor node deployed within this white (radio) zone. In the example of Figure 6.2, sensor node S1 can communicate with S2 but not with S3. That is why we can draw a link between S1 and S2 but not between the other pairs. The gray zone corresponds to the detection zone. It depends on the task the sensor node has been conceived for: to measure temperature, humidity, CO_2, ultrasound, or to detect a motion, a position, an intrusion etc. If, in addition, we assume that T1 and T2 are two moving objects, like cars or persons, and the sensor unit of S1 is a motion sensor, then S1 can detect T1 but not T2.

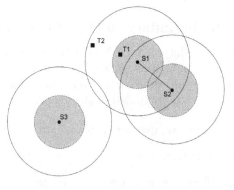

Figure 6.2: Three wireless sensor nodes, their radio range and detection zones.

Finding the boundary nodes of a WSN has two advantages: the first is to save energy in the nodes that are not on the boundary, and the second is to be able to use the other nodes in case a node on the boundary fails (due to loss of energy, interference, breakdown) or is getting hacked. To find such a boundary, we can use the LPCN algorithm, which in the form of Algorithm 5, requires a sink node for its execution. As shown by Figure 6.3(a), each node of the network sends its coordinates to the sink, which will run the LPCN algorithm and, upon termination, will send to each sensor node the information whether it is a boundary node or not (see Figure 6.3(b)).

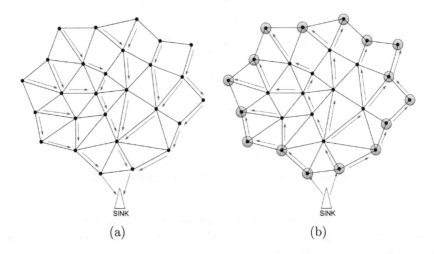

(a)   (b)

Figure 6.3: Boundary nodes of a WSN as determined by the centralized LPCN algorithm.

Another possibility is to use the D-LPCN algorithm (Algorithm 20) to be executed by each node of the network, independently. This algorithm has to be started from a preselected node that can be fixed manually or automatically using one of the leader election algorithms presented in Section 4.6 of Chapter 4. It is also possible to use D-RRLPCN as presented in Section 4.7.2 of Chapter 4. It has been shown in [100] that the centralized LPCN is more energy consuming than the distributed D-LCPN, because in the centralized version, all nodes must send and route their coordinates to the sink, and once the boundary nodes are calculated, the sink has to inform each sensor node whether it is on the boundary or not. The distributed version, however, requires only the boundary nodes and their neighbors, whereas the other nodes will not receive or send any messages. Figure 6.4 shows an example of a WSN in which the light-gray nodes represent the boundary nodes of which only the dark-gray neighbors participate in the process of finding the boundary nodes by executing the D-LPCN algorithm.

For this second version, since the nodes are independent, the starting node, which must be the one having the minimum x-coordinate, is determined manually. The main problem in reality is when the starting node cannot start the algorithm in case it is hacked or faulty, for example. That is why it is necessary to select this node automatically. To do this, we have presented in Chapter 4 some leader election algorithms that can be used, such as: LOGO,

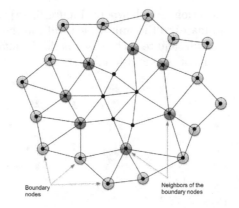

Figure 6.4: Boundary nodes of a WSN and their neighbors as the communicating nodes of the distributed D-LPCN algorithm.

BrOGO, DoTRo, WBS, etc. These algorithms are fault-tolerant because they allow to start the process of the D-LPCN even when there exist some faulty sensor nodes in the network.

Once the D-LPCN is executed, it is possible to be confronted with a situation where faulty nodes can arise accidentally. A method dealing with this problem will be presented in the next section.

6.2   Boundary node failure detection and reconfiguration

Once the boundary nodes are determined in a WSN, a node may be broken or getting hacked after a certain time. This kind of situation has been extensively studied in Lalem et al. [77]. If such a situation occurs, it is necessary to launch three major operations:

1. Detect the faulty node,

2. Inform the base station,

3. Reconfigure the boundary nodes, which means that a new boundary must be determined.

In some critical applications, informing the base station can be followed by an alarm triggering. The reconfiguration of the boundary can be done just by re-executing the D-LPCN algorithm. If a node is broken, it will not be considered for the new boundary determination. Figure 6.5 shows a network formed by 3 rings which is a good configuration for the execution of D-LPCN. Figure 6.5(a) shows in light-gray the resulting boundary nodes. Figure 6.5(b) shows an example where D-LPCN is re-executed after the detection of some faulty nodes in the first ring, and Figure 6.5(c) shows an example where many faulty nodes create a gap in the ring. As we can see, the D-LPCN algorithm continues to work even in such a situation. We call this process *Reconfiguration*.

The process of detecting faulty nodes is based on D-LPCN's specificity to determine the next boundary node among the neighbors of the current one. This means that each boundary node knows its subsequent node on the polygon hull. It is this property that allows one to run the *test of presence* procedure in which each boundary node will send a message to the subsequent one to ask for its presence. If this node is not faulty, it will answer the asking node that it is still active. We call this process *Boundary Node Failure Detection (BNFD)*.

The main steps of the BNFD process are presented in the flowchart of Figure 6.6.

First, the D-LPCN algorithm will be executed on a given WSN. As an example, Figure 6.7 shows a WSN with 13 sensor nodes and its boundary given by $S1$, $S2$, $S3$, $S4$, $S5$, $S6$ and $S7$. During the execution of this algorithm and in each iteration, every boundary node stores locally the identifier *id* of that neighbor, which will be selected as the next boundary node. Once the polygon hull of the network determined, every boundary node starts sending periodically a test message A *"Are you there?"* in order to test the presence of its boundary neighbor, as shown in Figure 6.7(a). Note that the A-messages can either be sent by all nodes at the same time or in a sequential way in order to avoid collisions, as described by the flowchart of Figure 6.6. Once an A-message is sent, the transmitter will wait for limited time for an answer message B *"Yes, I am"* from its next boundary neighbor, represented by gray arrows in Figures 6.7(b) and (d). If the neighbor is failing, it cannot answer. Therefore, once the limited time is over and no answer has been received from the neighbor (see Figure 6.7(c)), this neighbor will be declared as faulty (see Figure 6.7(d)). In this case, the process of reconfiguration will be started, i.e., the D-LPCN algorithm will be re-executed while an alarm will be triggered.

Algorithm 23 resumes the pseudocode of the *test of presence* procedure, where each node sends a message 'A' to its next boundary neighbor *nid* (cf. lines 2 and 3) once the boundary nodes are determined. Then each node will wait a given time t_1 for an answer message 'B' from its next neighbor (cf. line 4). If the node does not receive any message during this period, then this neighbor will be declared as a failure node (cf. lines 5 and 6) case in which the D-LPCN algorithm will be re-executed in order to find the new boundary nodes and an alarm will be triggered. Otherwise, if a node receives a message 'A' from its previous neighbor *pid* then it has to answer by a message 'B' in order to confirm that it is not a failure node (cf. lines 10, 11 and 12). Receiving the message 'B' from its next neighbor *nid* means that the node *nid* is not

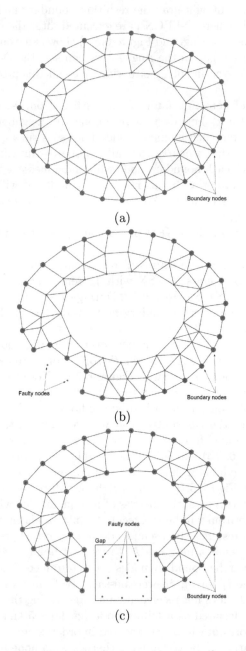

Figure 6.5: Examples for boundary reconfiguration.

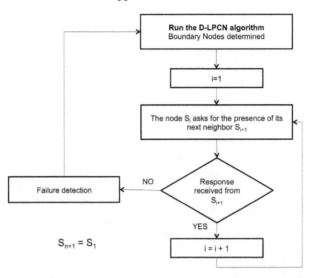

Figure 6.6: Faulty boundary node detection using the test of presence.

faulty (cf. lines 14 and 15). This algorithm is repeated periodically every t_2 seconds.

6.3 Finding voids and gaps in WSNs

A *void* in a WSN is a part of the space not covered by the union of the nodes' sensing areas. Such a part can cause serious problems for example in military applications where this area can be a source of intrusion or attacks. Before determining a void, we need first to determine an interior polygon as it is explained in Section 1.11.3, where two characteristics allow to determine whether a found polygon is interior or exterior.

Knowing the type of a polygon can be useful in real applications, like in WSNs, where we can use it to determine whether we have a *gap* or a *void* in a network.

A *gap* is simply the surface bounded by an interior polygon delimited by a set of sensor nodes in terms of radio communication, that exceeds a given threshold, and a *void* can be defined as an interior polygon containing an area which is not covered in the gap by the detection zones of its sensor nodes assuming that the detection radius is smaller than the radio communication range. In case of gaps, we need to determine the polygon formed by a given set of sensor nodes, which must be an interior polygon. In case of a void, we need to determine, in addition, whether the zone of this polygon is not totally

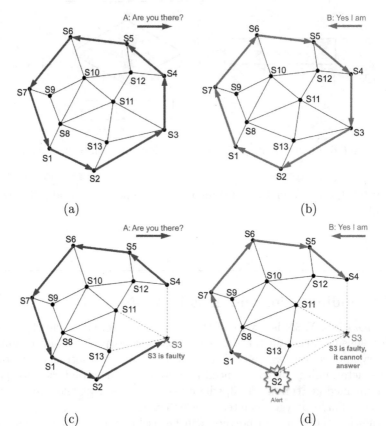

Figure 6.7: Boundary monitoring procedure.

Algorithm 23: Test of presence.

```
 1  while (true) do
 2  |   p = cid+"|"+"A";
 3  |   send(p, nid);
 4  |   n = read(t₁);
 5  |   if (n=="") then
 6  |   |   A failure is detected: run the D-LPCN algorithm;
 7  |   |   break();
 8  |   else
 9  |   |   type = read();
10  |   |   if (type==A) then
11  |   |   |   p = cid+"|"+"B";
12  |   |   |   send(p, pid);
13  |   |   end
14  |   |   if (type==B) then
15  |   |   |   No failure detected: do nothing;
16  |   |   end
17  |   end
18  |   wait(t₂);
19  end
```

---

covered by the detection areas of the sensor nodes forming it. Figure 6.8(a) shows in gray the polygon found by D-LPCN when starting from node A. This polygon is an interior polygon. If we look at the detection zones of its nodes, i.e., the light-gray area in Figure 6.8(c), we observe that there is a dark-grey zone, which is un-covered. However, if we start from node C, we will find a triangle, shown in gray in Figure 6.8(b), which is not a void because, from Figure 6.8(d), the detection zones of the three sensor nodes completely cover the triangle.

A different situation arises when we start from node B, which is a boundary node. We obtain the exterior polygon shown in Figure 6.8(e) which is not a void because it represents the boundary of the whole network. This can be verified by calculating the sum of the angles using Equation (1.4).

Now, if we assume that there is no interior connection between nodes, as shown by Figure 6.8(f), and if we start D-LPCN from node E we will obtain the same polygon as in Figure 6.8(e), but it is now an interior polygon and the formula of Equation (1.4) will be verified. As we can see, the dark-gray zone is not covered by the detection zones of the nodes of the polygon.

Another problem in detecting voids in WSNs is the calculation of the detection area's surface. In other words, once the boundary nodes are determined, how can we know that there is an uncovered area? A simple answer is by calculating the surface of the obtained polygon hull. We can fix a certain threshold and if the surface is greater than this threshold, we will consider that a void is detected.

In the following, we will present Algorithm 24, presented by Bezoui et al. [13], the new version of Algorithm 20 (D-LPCN) to integrate the code allowing to determine the nature of the found polygon (exterior or interior). We have added the line 15 to calculate the first angle. Since the first angle is based on a fictitious point situated to the left of the starting node, only a part of the first angle is calculated. The variable f_angle allows to calculate the other part which is the angle situated between the fictitious node and the neighbor node forming the maximum polar angle. This node is also the one forming the minimum anti-polar angle between the fictitious node and the neighbors. The variable phi_max is calculated in lines 33 to 35. The variable t_angle of line 16 allows to calculate the sum of the angles of the polygon. Each node will receive this value from its previous neighbor (p_angle), and then add it to the value of its own angle. The same procedure is used for the number of visited nodes nbr_bn, which is to be compared with the value given by Equation (1.4). Once these values are updated, they will be sent to the next neighbor n_id in line 19. Lines 39 to 45 concern only the starting node where the received sum of the angles of the polygon can be correct or not. If it is correct, then the obtained polygon is interior, otherwise it is exterior.

We will conclude by presenting Algorithm 25, presented by Bounceur et al. [25], which is based on the global minimum of the found polygon. As in the previous case, it is the D-LPCN algorithm which is modified. In line 52, each node will test whether its x-coordinate is smaller than the one received from its neighbor (line 49). This will help to determine whether the variable is_min is true or false (lines 52 to 55), and also in lines 15 and 16 to determine which angle will be sent, the received one (line 49) or the currently calculated one (line 28). As soon as the values of min_x and min_angle have been determined, they will be sent to the next neighbor (line 20).

6.4   Cluster finding and shape reconstruction

In this section, we will show how to determine the different clusters of a two-dimensional dataset. The main idea for this is to first connect the points using the α-shape algorithm presented in Section 2.3 or simply using the UDG (*unit disque graph*) model [36] as in WSNs. That is to say, we assume that each point has a communication range of a given radius and each further point situated within this range will be connected to it. Figure 6.9 shows an example of a point set the elements of which are connected on the basis of a circular range.

Once the points are connected, we can execute LPCN to find the boundary nodes of the first cluster containing the point with the smallest x-coordinate.

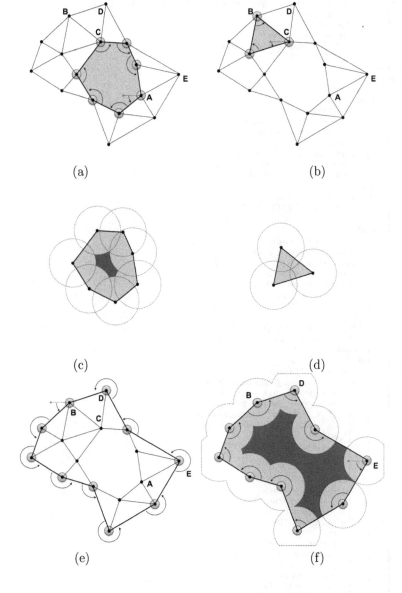

Figure 6.8: Examples of a void in a WSN.

Algorithm 24: Type of polygon determined by angles.

```
 1  boundary = false; phi_min = 10; phi_max = -10;
 2  c_id = getId(); c_coord = getCoord();
 3  boundary_set = ∅; i=0; n = getNumberOfNeighbors(); selected = false;
 4  first_node = any node;
 5  if (first_node) then
 6  |    boundary = true;
 7  |    p_coord = (c_coord.x−1, c_coord.y);
 8  |    send(c_id+"|"+"AC", *);
 9  end
10  repeat
11  |    id = read();
12  |    type = read();
13  |    if (i==n) then
14  |    |    boundary_set = boundary_set ∪ {c_id};
15  |    |    f_angle =6.28319−phi_max;
16  |    |    t_angle = phi_min + p_angle;
17  |    |    send(c_id+"|"+"SN"+"|"+
18  |    |    c_coord+"|"+boundary_set, n_id);
19  |    |    send(t_angle+"|"+nbr_bn,n_id);
20  |    end
21  |    if (type=="AC") then
22  |    |    send(c_id+"|"+"CS"+"|"+
23  |    |    c_coord, id);
24  |    end
25  |    if (type=="CS") then
26  |    |    n_coord = read(); i=i+1;
27  |    |    phi = angleWI(p_coord,c_coord,
28  |    |    n_coord, boundary_set);
29  |    |    if (phi<phi_min) then
30  |    |    |    phi_min = phi;
31  |    |    |    n_id = id;
32  |    |    end
33  |    |    if (phi>phi_max) then
34  |    |    |    phi_max = phi;
35  |    |    end
36  |    end
37  |    if (type=="SN") then
38  |    |    if (selected and first_node) then
39  |    |    |    t_angle = f_angle+p_angle;
40  |    |    |    ref_angle=(nbr_bn−2)×3.141595;
41  |    |    |    if (t_angle == ref_angle) then
42  |    |    |    |    print "INTERIOR";
43  |    |    |    else
44  |    |    |    |    print "EXTERIOR";
45  |    |    |    end
46  |    |    |    stop();
47  |    |    else
48  |    |    |    selected = true;
49  |    |    |    boundary=true;
50  |    |    |    phi_min=10;
51  |    |    |    phi_max=-10;
52  |    |    |    i=0;
53  |    |    |    p_coord = read();
54  |    |    |    boundary_set = read();
55  |    |    |    p_angle=read();
56  |    |    |    nbr_bn=read();
57  |    |    |    nbr_bn = nbr_bn + 1;
58  |    |    |    send(c_id+"|"+"AC", *);
59  |    |    end
60  |    end
61  until false;
```

Algorithm 25: Type of polygon determined by global minimum node.

```
1  boundary = false; phi_min = 10; phi_max = -10;
2  c_id = getId(); c_coord = getCoord();
3  boundary_set = ∅; i=0; n = getNumberOfNeighbors(); selected = false;
4  first_node = any node;
5  if (first_node) then
6      boundary = true;
7      p_coord = (c_coord.x−1, c_coord.y);
8      send(c_id+"|"+"AC", *);
9  end
10 repeat
11     id = read();
12     type = read();
13     if (i==n) then
14         boundary_set = boundary_set ∪ {c_id};
15         if (is_min) then
16             | min_angle = phi_min;
17         end
18         send(c_id+"|"+"SN"+"|"+
19         c_coord+"|"+boundary_set, n_id);
20         send(min_x+"|"+min_angle, n_id);
21     end
22     if (type=="AC") then
23         send(c_id+"|"+"CS"+"|"+
24         c_coord, id);
25     end
26     if (type=="CS") then
27         n_coord = read(); i=i+1;
28         phi = angleWI(p_coord, c_coord, n_coord, boundary_set);
29         if (phi<phi_min) then
30             phi_min = phi;
31             n_id = id;
32         end
33     end
34     if (type=="SN") then
35         if (selected and first=_node) then
36             if (min_angle ≤ 3.141595) then
37                 | print "INTERIOR";
38             else
39                 | print "EXTERIOR";
40             end
41             stop();
42         else
43             selected = true;
44             boundary = true;
45             phi_min = 10;
46             i=0;
47             p_coord = read();
48             boundary_set = read();
49             min_x = read();
50             min_angle = read();
51             is_min = false;
52             if (c_coord.x < min_x) then
53                 is_min = true;
54                 min_x = c_coord.x;
55             end
56             send(c_id+"|"+"AC", *);
57         end
58     end
59 until false;
```

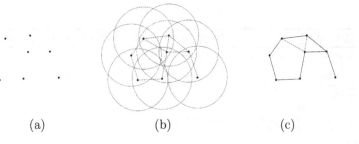

(a)   (b)   (c)

Figure 6.9: A point set: (a) unconnected, (b) connected using a UDG model, (c) with its boundary.

This cluster is defined by all those points lying inside the found polygon hull. To find a second cluster, we remove from the dataset all points of the first cluster and then apply the same procedure to the remaining points. This process is renewed until all clusters have been found. Our procedure can extract complex clusters from a dataset with the condition that the distance between each couple of clusters must be greater than the radius fixed to connect the points. Figure 6.10 shows an example of a two-dimensional dataset (see Figure 6.10(a)), the connection between points (Figure 6.10(b)) and the clusters found (see Figure 6.10(c)). A similar approach can be used for shape reconstruction. Since the complexity of the LPCN algorithm depends on the number of boundary points, our approach can be applied to huge datasets, which makes it very useful in the context of Big Data.

This algorithm gives good results only when the clusters are well defined, i.e., do not contain subclusters or gaps. In the case of ring graphs, only the first boundary is found and not the one lying inside the ring nor the subcluster contained in the interior (see Figure 6.11(a)). To overcome this problem, we propose to determine all the nodes that are connected to the boundary nodes found in the first run, then their neighbors and the neighbors of the neighbors and so on. To do this, we can start from any node of the boundary to visit all the nodes of the ring using the DFS process or by visiting the nodes of the ring layer by layer, i.e., visit the first-hop neighbors of the boundary nodes, and then the second-hop neighbors until all the nodes of the ring have been reached. For this, we simply modify line 12 of Algorithm 26 to obtain Algorithm 27 whose application to our example produces the result shown in Figure 6.11(b).

To determine just the boundary nodes of the ring, we can start either from a local minimum situated in the last layer of the cluster which can be determined as explained previously. If we start the LPCN algorithm from this node, we will find an interior polygon which will represent the inner polygon of the ring of the first cluster as shown by Figure 6.11(c). The combination

(a)

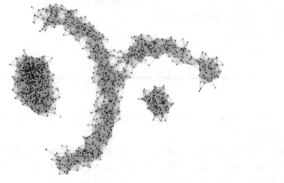

(b)

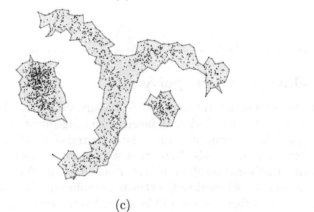

(c)

Figure 6.10: Finding clusters.

Algorithm 26: Cluster extraction (Version 1).

Data: S: set of n Points $v_i, i = 1..n$; r: real value (threshold)
Result: P set of Clusters

1 $P = \emptyset$;
2 Transform the set S into a graph G, where each couple of points having a Euclidean distance smaller than r will be connected $k \leftarrow 1$;
3 **while** *($\exists$ non marked point in S)* **do**
4 $\quad$ $S' \leftarrow$ non marked points of S;
5 $\quad$ $G' \leftarrow$ the subgraph of G formed by the points of S';
6 $\quad$ $x_{min} \leftarrow \underset{v_i=(x_i,y_i)\in G'}{\mathrm{argmin}}\{x_i\}$;
7 $\quad$ Run the LPCN algorithm on G' starting from x_{min};
8 $\quad$ Let P_k be the polygon found;
9 $\quad$ Mark all the points of P_k;
10 $\quad$ $W \leftarrow \{v|v \in G'$ & v is inside the polygon $P_k\}$;
11 $\quad$ Mark all the points of W;
12 $\quad$ $P \leftarrow P \cup P_k$;
13 $\quad$ $k \leftarrow k + 1$;
14 **end**

---

of this method with Algorithm 27 allows to find all the clusters as shown by Figure 6.11(d).

If there is no local minimum, then it is not possible to use this method in general graphs but still in plane graphs where we can take the node in the last layer having the maximum x-coordinate and start the LPCN algorithm from it.

6.5   Image contour polygon

In this section, we will show how to find and draw the polygon representing the contours of a zone within an image. The image must be binary (black and white), and in the opposite case, it must be transformed into a binary image using intensity thresholds. Note, that we are not trying to do segmentation, i.e., extract information about objects from an image. For our case, we assume that the object itself is already extracted graphically since the image must be binary. We will then use the LPCN algorithm to transform a zone of interest of the image into one or several polygons. It should also be possible to combine LPCN with another algorithm that allows to characterize such a zone.

For illustration, let us consider the extraction of the gray zone of the image of Figure 6.12(a). First, we transform the region containing this zone into a

---

Algorithm 27: Cluster extraction (Version 2).

Data: S: set of n Points $v_i, i = 1..n$; r: real value (threshold)
Result: P set of Polygons
1 $P = \emptyset$;
2 Transform the set S into a graph G, where each couple of points having a Euclidean distance smaller than r will be connected;
3 $k \leftarrow 1$;
4 **while** *($\exists$ non marked point in S)* **do**
5 $\quad$ $S' \leftarrow$ non marked points of S;
6 $\quad$ $G' \leftarrow$ the subgraph of G formed by the points of S';
7 $\quad$ $x_{min} \leftarrow \underset{v_i = (x_i, y_i) \in G'}{\operatorname{argmin}}\{x_i\}$;
8 $\quad$ Run the LPCN algorithm on G' starting from x_{min};
9 $\quad$ Let P_k be the polygon found;
10 $\quad$ Mark all the points of P_k;
11 $\quad$ *found* $\leftarrow$ *true*;
12 $\quad$ **while** *(found)* **do**
13 $\quad\quad$ *found* $\leftarrow$ *false*;
14 $\quad\quad$ **for** *(each point p of P_k)* **do**
15 $\quad\quad\quad$ $W \leftarrow \{v | v \in G'$ & v is a neighbor of $p\}$;
16 $\quad\quad\quad$ *found* $\leftarrow$ *true*;
17 $\quad\quad$ **end**
18 $\quad$ **end**
19 $\quad$ Mark all the points of W;
20 $\quad$ $P \leftarrow P \cup P_k$;
21 $\quad$ $k \leftarrow k + 1$;
22 **end**

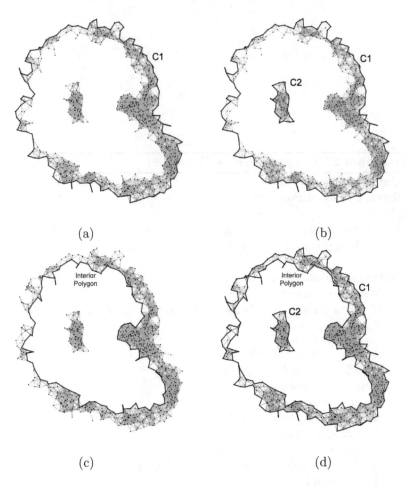

Figure 6.11: Finding clusters.

matrix of pixels, which can be modeled as a connected Euclidean graph, as shown by Figures 6.12(b) and (c).

It turns out that the obtained graph does not contain any of the critical cases (analysed and resolved in [76]) for an application of LPCN1. Now, if we are interested in a simple extraction based on the intensity of pixels, we will consider only the light-gray points of the graph, as shown by Figure 6.12(d), which correspond to the gray pixels of the image. For this graph, we are able to run the LPCN algorithm to obtain its boundary points (see Figure 6.12(e)). Figures 6.12(f) and (g) show that these points represent the contour in light-gray of the gray zone in the original image. As a more realistic example, Figure 6.13 shows the contour of a tumor extracted by using the LPCN algorithm. This picture represents a tumor visualized by magnetic resonance imaging at the University Hospital of Brest.

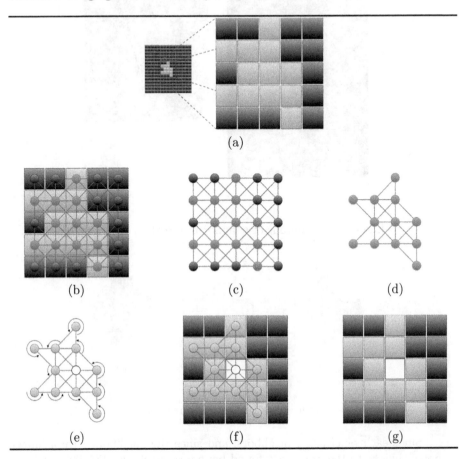

Figure 6.12: Contour of a zone of interest.

Finding the polygon hull of a set of pixels representing the contour of a zone of interest has many advantages, such as:

- Reducing the storage size of the files by saving only polygons instead of whole images,

- Reconstructing the initial image from the polygons as shown by Figure 6.14,

- Drawing the contours of the zone of interest to improve a drawing obtained manually, as shown by Figure 6.15, where Figure 6.15(a) shows the initial zone of interest, Figure 6.15(b) shows the polygon drawn manually and Figure 6.15(c) the one found by LPCN.

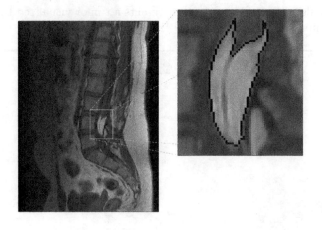

Figure 6.13: Contour of a tumor in a real medical image.

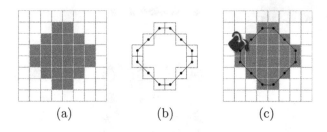

(a)   (b)   (c)

Figure 6.14: Image contour polygon.

Since each pixel has 8 neighbors, and since the neighbors form a known polar angle with the previously found boundary pixel, i.e., 45°, 90°, 135°, ..., 360°, we can accelerate the LPCN algorithm in this case by replacing the calculation of angles by just visiting the neighbors following a polar rotation and taking the first pixel encountered from the previously found boundary pixel. To explain this in more detail, we will use the example of Figure 6.16, where we try to find the boundary pixels of the black pixels using the LPCN algorithm. First, we start from the pixel with the minimum x-coordinate. There are two such pixels and we are free to choose. Let us take the one which is on bottom. This pixel will be the first boundary pixel of the black zone. It is represented in gray color with a dot in Figure 6.16(b). In the first iteration, the fictitious pixel considered is its left neighbor (pixel 8 of Figure 6.16(b)). From this pixel, we number the neighbor pixels of the starting pixel from 1 (for the pixel situated after the fictitious pixel in polar order) to 8. We visit each pixel in ascending order of these numbers. The first encountered black pixel will be the next boundary pixel. In this example, the next boundary

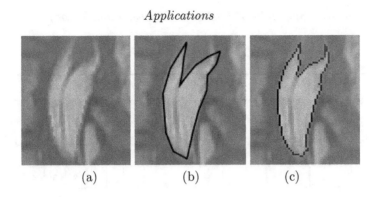

(a) (b) (c)

Figure 6.15: Image contour drawing.

pixel is the pixel numbered 3 in Figure 6.16(b), which is shown in gray with a dot in Figure 6.16(c). In the same way, we number the neighbor pixels of the current boundary pixel (in gray with a dot) from 1 (for the pixel situated after it in polar order) to 8. Note, that in the next iteration, the previously found boundary pixel will have the number 8. We continue this process, as shown in Figures 6.16(d), (e) and (f), until the stopping condition is reached, i.e., pixel 6 of Figure 6.16(f) visits the next boundary node for a second time. If we continue to apply LPCN, we get the situation of Figure 6.16(b), where node 3 is visited again. So we stop the algorithm and the boundary nodes found together with the corresponding polygon hull are shown in Figure 6.17. For this case we may rebaptize the LPCN algorithm as Least Polar-angle Connected Pixel (LPCP).

Another application where we can use the LPCP algorithm is to find the clusters in an image as presented in Section 6.4, but where we use a graph of pixels instead of a graph of nodes. This can be useful in the case of fingerprints where we save the polygons of the different black parts instead of the whole image. Thereafter, we need just to compare polygons between them by using existing methods such as shape matching techniques [10, 114]. We propose a simple method for this by comparing two polygons using their difference or substraction. To better explain this method, let us consider one part of two fingerprints that are taken from the same person, a reference one and another one which we want to compare it with. It is preferable to use the same conditions and the same material to get the fingerprints. Otherwise, some operations are needed before comparison. Here, we assume that the second part is smaller than the reference part. We start to determine the polygon hulls P1 and P2 of each part using the image contour polygon method presented in the previous section. Figure 6.18 shows the polygons obtained for each part.

Since in this example the polygon P2 is smaller than the reference polygon P1, we need just to zoom to get the same scale based on the bounding box of each polygon, as shown by Figure 6.19.

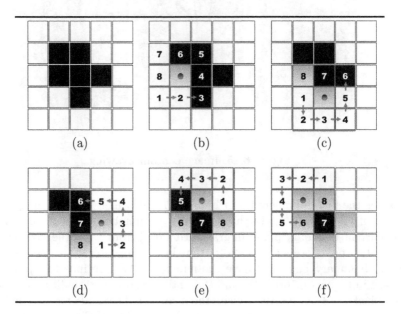

Figure 6.16: Contour of a set of pixels.

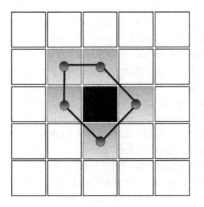

Figure 6.17: Contour of a set of pixels in form of a polygon.

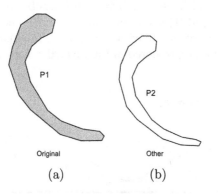

Figure 6.18: Two polygon hulls arising from fingerprints of the same person: (a) reference, (b) other.

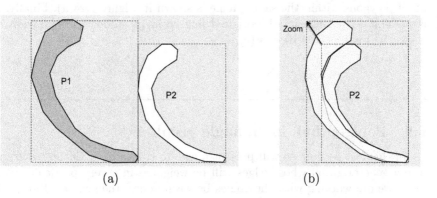

Figure 6.19: Zoom operation on the other polygon: (a) initial polygon P2, (b) the zoomed polygon P2.

Now, we superpose the two polygons P1 and P2 based on their bounding boxes, and we slightly shift, vertically or horizontally, the two polygons as shown by Figure 6.21, to obtain the smallest difference area between them. Figures 6.21(a), (c) and (e) show different shiftings between two polygons and Figures 6.21(b), (d) and (f) show the obtained difference area between them.

The surface values of the difference areas as a function of the shifting will be close to the function represented in Figure 6.20, where S denotes the surface of the polygon P1.

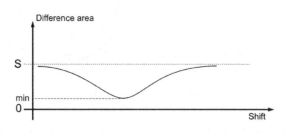

Figure 6.20: Difference values of two polygons with respect to their shifting.

We have applied this method to a real fingerprint example (Figure 6.22(a)). Figures 6.22(b) and (c) show the polygon of one part of this fingerprint as obtained by applying LPCP. Figure 6.22(d) shows the extracted polygon and, finally, Figure 6.22(e) shows the reconstructed part of the fingerprint using this polygon.

The polygons of the other parts are shown by Figure 6.23. The assembly of all polygons within the same image is shown in Figure 6.24(a). Finally, it is possible to reconstruct the original fingerprint by filling each polygon as shown by Figures 6.24(b) and (c).

6.6   Polygon hull in an angle graph

In this last section we will present a way to transform a Euclidean graph into a general graph whose edges will be weighted in a very particular way. Since we are working with the angles between nodes, the generated graph is called an *angle graph*. Such a graph has the same structure as the Euclidean graph except that the position of the nodes is not important. The edges will be replaced by pairs of arcs. The 'weight' of an arc depends on the order of visiting the nodes, where the notion of time or iteration must be taken into account. Thus, an arc (A, B) at iteration i depends on the node C just visited before A at iteration $i-1$. In the case where the previously visited node is not known, the angle will be calculated with respect to a fictitious node situated

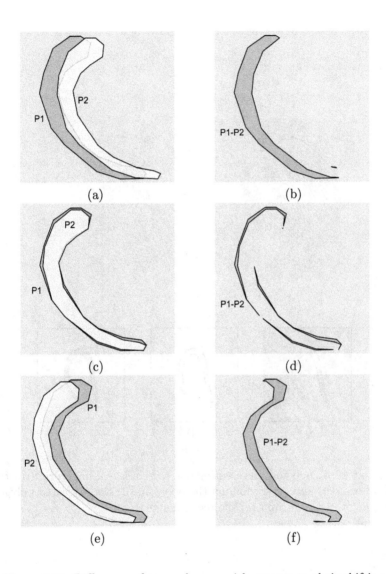

Figure 6.21: Difference of two polygons with respect to their shifting.

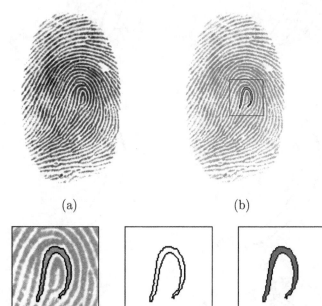

Figure 6.22: A polygon representing one part of a real fingerprint (a) the initial one, (b) and (c) a zoom on the polygon, (d) the extracted polygon and (e) the reconstructed part.

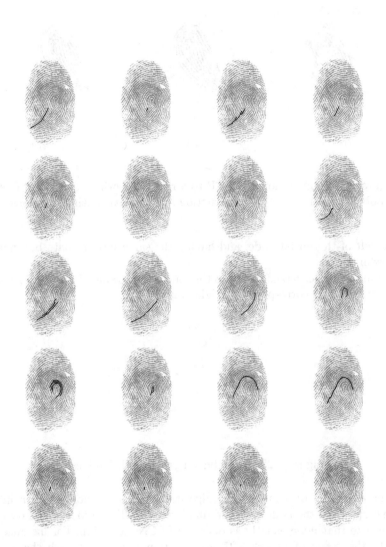

Figure 6.23: A list of polygons of different parts of a real fingerprint obtained by the LPCP algorithm.

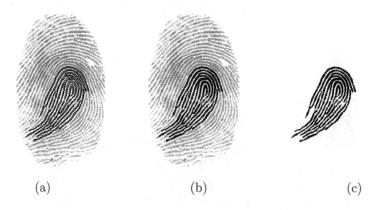

(a)   (b)   (c)

Figure 6.24: Application of LPCP to a real fingerprint: (a) the obtained polygons, (b) and (c) reconstruction of those parts using polygons.

at the left of the initial node, and having the same y-coordinate but smaller x-coordinate.

As an example, let us consider the Euclidean graph of Figure 6.25. Figure 6.26 shows its corresponding angle graph.

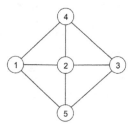

Figure 6.25: A Euclidean graph with 5 nodes.

The execution of a greedy-like algorithm to find a shortest path in the angle graph from the node with the smallest x-coordinate to a boundary node will lead to that node exactly as does the LPCN algorithm. For instance, let us take the graph of Figure 6.27. At iteration 1, we start with the node 1 having the smallest x-coordinate. In the angle graph, we only consider those arcs that have node 1 as initial node and where the previously visited node is the fictitious node 1', as shown in Figure 6.27(a). From this arc, we will choose the next as the one having the smallest cost of 135, arc which is connected to node 5. Thus, the next node to visit is node 5. From node 5 on, we continue to explore all the arcs with node 5 as initial node, as shown by Figure 6.27(b). The arc with the smallest cost of 270 is the arc (5, 3). We continue this way, as shown by Figures 6.27(c) and (d), until we visit node 1 for a second time,

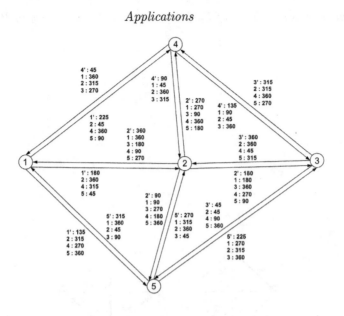

Figure 6.26: The corresponding angle graph.

which is the simplest stop condition for this case. As we can see, we finally come up with the boundary nodes of the graph of Figure 6.25.

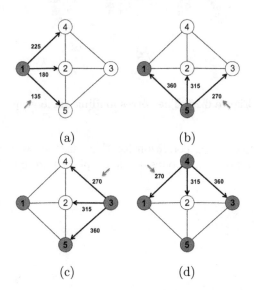

(a)   (b)

(c)   (d)

Figure 6.27: An example of a greedy-like algorithm to find a shortest path in an angle graph.

Bibliography

[1] Selim G Akl. Two remarks on a convex hull algorithm. *Information Processing Letters*, 8(2):108–109, 1979.

[2] Selim G Akl and Godfried T Toussaint. Efficient convex hull algorithms for pattern recognition applications. In *Proceedings of the 4th International Joint Conference on Pattern Recognition, Kyoto, Japan*, pages 483–487, 1978.

[3] Selim G Akl and Godfried T Toussaint. A fast convex hull algorithm. *Information Processing Letters*, 7(5):219–222, 1978.

[4] Li An, Qing-San Xiang, and Sofia Chavez. A fast implementation of the minimum spanning tree method for phase unwrapping. *IEEE Transactions on Medical Imaging*, 19(8):805–808, 2000.

[5] Kenneth R Anderson. A reevaluation of an efficient algorithm for determining the convex hull of a finite planar set. *Information Processing Letters*, 7(1):53–55, 1978.

[6] Alex M Andrew. Another efficient algorithm for convex hulls in two dimensions. *Information Processing Letters*, 9(5):216–219, 1979.

[7] R. Balakrishnan and K. Ranganathan. *A Textbook of Graph Theory*. Universitext (1979). Springer, 2000.

[8] Vladimir Batagelj and Matjaz Zaveršnik. Short cycle connectivity. *Discrete Mathematics*, 307(3):310–318, 2007.

[9] Luca Becchetti, Paolo Boldi, Carlos Castillo, and Aristides Gionis. Efficient semi-streaming algorithms for local triangle counting in massive graphs. In *Proceedings of the 14th ACM SIGKDD International Conference on Knowledge Discovery and Data Mining*, pages 16–24. ACM, 2008.

[10] Serge Belongie, Jitendra Malik, and Jan Puzicha. Shape matching and object recognition using shape contexts. *IEEE Transactions on Pattern Analysis and Machine Intelligence*, 24(4):509–522, 2002.

[11] Fausto Bernardini and Chandrajit L Bajaj. Sampling and reconstructing manifolds using alpha-shapes. *Computer Science Technical Reports, Paper 1350, Purdue University, USA*, 1997.

[12] Madani Bezoui, Ahcène Bounceur, Reinhardt Euler, and Mustapha Moulaï. A new distributed algorithm for finding dominating sets in IoT networks under multiple criteria. In *12th International Conference on Multiple Objective Programming and Goal Programming (MOPGP 2017)*, 2017.

[13] Madani Bezoui, Ahcène Bounceur, Loic Lagadec, Reinhardt Euler, Frank Singhoff, and Mohammad Hammoudeh. Detecting gaps and voids in WSNs and IoT networks: the angle-based method. In *International Conference on Future Networks and Distributed Systems (ICFNDS), 26-27 June, Amman, Jordan*, 2018.

[14] Cheng Bi, An Zhang, and Ying Wan. Time synchronisation in wireless sensor networks under energy-efficient spanning tree protocol. *International Journal of Automation and Logistics*, 2(3):218–233, 2016.

[15] Jean-Daniel Boissonnat. Geometric structures for three-dimensional shape representation. *ACM Transactions on Graphics (TOG)*, 3(4):266–286, 1984.

[16] Béla Bollobás. *Modern Graph Theory*, volume 184. Springer Science & Business Media, 2013.

[17] Ahcène Bounceur. Cupcarbon: a new platform for designing and simulating smart-city and IoT wireless sensor networks (SCI-WSN). In *Proceedings of the International Conference on Internet of things and Cloud Computing*, page 1. ACM, 2016.

[18] Ahcène Bounceur. Finding the boundary nodes of a WSN using the D-LPCN algorithm and its simulation under cupcarbon. In *1st EAI International Conference on Future Internet Technologies and Trends (ICFITT 2017), August 31-September 2, 2017, Surat, India*, 2017.

[19] Ahcène Bounceur. Cupcarbon simulator, 2018.

[20] Ahcène Bounceur. *CupCarbon User Guide*. 2018.

[21] Ahcène Bounceur, Madani Bezoui, Reinhardt Euler, Nabil Kadjouh, and Farid Lalem. BROGO: A new low energy consumption algorithm for leader election in WSNs. In *Developments in E-Systems Engineering (DeSE 2017), 14-16 June, Paris, France*, 2017.

[22] Ahcène Bounceur, Madani Bezoui, Reinhardt Euler, and Farid Lalem. A Wait-Before-Starting Algorithm for Fast, Fault-Tolerant and Low Energy Leader Election in WSNs Dedicated to Smart-Cities and IoT. In *IEEE Sensors, Glasgow, United Kingdom*, 2017.

[23] Ahcène Bounceur, Madani Bezoui, Reinhardt Euler, Farid Lalem, and Massinissa Lounis. A revised brogo algorithm for leader election in

wireless sensor and IoT networks. In *IEEE Sensors, Glasgow, United Kingdom*, 2017.

[24] Ahcène Bounceur, Madani Bezoui, Reinhardt Euler, and Marc Sevaux. Finding the boundary nodes of a wireless sensor network without conditions on the starting node. In *28th European Conference on Operational Research (EURO 28)*, 2016.

[25] Ahcène Bounceur, Madani Bezoui, Loic Lagadec, Reinhardt Euler, Abdelkader Laouid, Mahamadou Traore, and Mounir Lallali. Detecting gaps and voids in WSNs and IoT networks: the minimum x-coordinate based method. In *International Conference on Future Networks and Distributed Systems (ICFNDS), 26-27 June, Amman, Jordan*, 2018.

[26] Ahcène Bounceur, Madani Bezoui, Massinissa Lounis, Reinhardt Euler, and Ciprian Teodorov. A New Dominating Tree Routing Algorithm for Efficient Leader Election in IoT Networks. In *Proceedings of IEEE Consumer Communications & Networking Conference (CCNC 2018), 12-15 January 2018, Las Vegas, USA*, 2018.

[27] Ahcène Bounceur, Madani Bezoui, Umber Noreen, Reinhardt Euler, Farid Lalem, Mohammad Hammoudeh, and Sohail Jabbar. LOGO: A new distributed leader election algorithm in WSNs with low energy consumption. In *International Conference on Future Internet Technologies and Trends ICFITT*, pages 1–16. Springer, 2017.

[28] Ahcène Bounceur, Reinhardt Euler, Ali Benzerbadj, Farid Lalem, Massinissa Saoudi, and Marc Sevaux. Finding the polygon hull in wireless sensor networks. In *27th European Conference on Operational Research (EURO 27)*, 2015.

[29] Ahcène Bounceur, Olivier Marc, Massinissa Lounis, Julien Soler, Laurent Clavier, Pierre Combeau, Rodolphe Vauzelle, Loïc Lagadec, Reinhardt Euler, Madani Bezoui, and Pietro Manzoni. Cupcarbon-Lab: An IoT emulator. In *IEEE Consumer Communications & Networking Conference, 12-15 January 2018, Las Vegas, USA*, 2018.

[30] Alex Bykat. Convex hull of a finite set of points in two dimensions. *Information Processing Letters*, 7(6):296–298, 1978.

[31] Constantin Carathéodory. Über den Variabilitätsbereich der Fourier'schen Konstanten von positiven harmonischen Funktionen. *Rendiconti del Circolo Matematico di Palermo (1884-1940)*, 32(1):193–217, 1911.

[32] Timothy M Chan. Optimal output-sensitive convex hull algorithms in two and three dimensions. *Discrete & Computational Geometry*, 16(4):361–368, 1996.

[33] A Ray Chaudhuri, Bidyut Baran Chaudhuri, and Swapan K Parui. A novel approach to computation of the shape of a dot pattern and extraction of its perceptual border. *Computer Vision and Image Understanding*, 68(3):257–275, 1997.

[34] Gilbert Chen, Joel Branch, Michael Pflug, Lijuan Zhu, and Boleslaw Szymanski. Sense: a wireless sensor network simulator. In *Advances in Pervasive Computing and Networking*, pages 249–267. Springer, 2005.

[35] L Paul Chew. There is a planar graph almost as good as the complete graph. *Dartmouth College, Hanover, NH03755*, 1986.

[36] Brent N. Clark, Charles J. Colbourn, and David S. Johnson. Unit disk graphs. *Discrete Mathematics*, 86(1):165–177, 1990.

[37] John Clark and Derek Allan Holton. *A first look at graph theory*. World Scientific, 1991.

[38] Ryan R Curtin and Marcus Edel. Designing and building the mlpack open-source machine learning library. *arXiv preprint: 1708.05279*, 2017.

[39] Mark De Berg, Marc Van Kreveld, Mark Overmars, and Otfried Cheong Schwarzkopf. Computational geometry. In *Computational Geometry*, pages 1–17. Springer, 2000.

[40] Boris Delaunay. Sur la sphere vide. *Izv. Akad. Nauk SSSR, Otdelenie Matematicheskii i Estestvennyka Nauk*, 7(793-800):1–2, 1934.

[41] Erik Demaine, Joseph Mitchell, and Joseph O'Rourke. The open problems project. *http://maven.smith.edu/ orourke/TOPP/ (see problem 57, Chromatic Number of the Plane)*, 2009.

[42] William F Eddy. A new convex hull algorithm for planar sets. *ACM Transactions on Mathematical Software (TOMS)*, 3(4):398–403, 1977.

[43] H. Edelsbrunner, D. Kirkpatrick, and R. Seidel. On the shape of a set of points in the plane. *IEEE Transactions on Information Theory*, 29(4):551–559, Jul 1983.

[44] Herbert Edelsbrunner. Alpha shapes-a survey. *Tessellations in the Sciences*, 27:1–25, 2010.

[45] Herbert Edelsbrunner and Ernst P Mücke. Three-dimensional alpha shapes. *ACM Transactions on Graphics (TOG)*, 13(1):43–72, 1994.

[46] István Fáry. On straight line representation of planar graphs. *Acta. Sci. Math. Szeged*, 11:229–233, 1948.

[47] Marwan Fayed and Hussein T Mouftah. Localised alpha-shape computations for boundary recognition in sensor networks. *Ad Hoc Networks*, 7(6):1259–1269, 2009.

[48] Margaret M Fleck. The topology of boundaries. *Artificial Intelligence*, 80(1):1–26, 1996.

[49] Gautam Garai and BB Chaudhuri. A split and merge procedure for polygonal border detection of dot pattern. *Image and Vision Computing*, 17(1):75–82, 1999.

[50] Bernd Gärtner and Michael Hoffmann. Computational geometry lecture notes 1. *HS*, 2013.

[51] Leonardo Gomes, Olga Regina Pereira Bellon, and Luciano Silva. 3D reconstruction methods for digital preservation of cultural heritage: A survey. *Pattern Recognition Letters*, 50:3–14, 2014.

[52] Hongyu Gong, Luoyi Fu, Xinzhe Fu, Lutian Zhao, Kainan Wang, and Xinbing Wang. Distributed multicast tree construction in wireless sensor networks. *IEEE Transactions on Information Theory*, 63(1):280–296, 2017.

[53] M Gopi, Shankar Krishnan, and Cláudio T Silva. Surface reconstruction based on lower dimensional localized Delaunay triangulation. In *Computer Graphics Forum*, volume 19, pages 467–478. Wiley Online Library, 2000.

[54] Ronald L Graham. An efficient algorithm for determining the convex hull of a finite planar set. *Information Processing Letters*, 1(4):132–133, 1972.

[55] Ronald L Graham and F Frances Yao. Finding the convex hull of a simple polygon. *Journal of Algorithms*, 4(4):324–331, 1983.

[56] David Gries and Ivan Stojmenović. A note on Graham's convex hull algorithm. *Information Processing Letters*, 25(5):323–327, 1987.

[57] Zhenqun Guan, Chao Song, and Yuanxian Gu. The boundary recovery and sliver elimination algorithms of three-dimensional constrained Delaunay triangulation. *International Journal for Numerical Methods in Engineering*, 68(2):192–209, 2006.

[58] Frank Harary and Helene J Kommel. Matrix measures for transitivity and balance. *Journal of Mathematical Sociology*, 6(2):199–210, 1979.

[59] John Hershberger and Subhash Suri. Applications of a semi-dynamic convex hull algorithm. *BIT Numerical Mathematics*, 32(2):249–267, 1992.

[60] Hiroshi Hirai. Tight spans of distances and the dual fractionality of undirected multiflow problems. *Journal of Combinatorial Theory, Series B*, 99(6):843–868, 2009.

[61] Christian Hirsch, David Neuhäuser, Catherine Gloaguen, and Volker Schmidt. Asymptotic properties of Euclidean shortest-path trees in random geometric graphs. *Statistics & Probability Letters*, 107:122–130, 2015.

[62] Petter Holme and Beom Jun Kim. Growing scale-free networks with tunable clustering. *Physical Review E*, 65(2):026107, 2002.

[63] Chi-Fu Huang and Yu-Chee Tseng. The coverage problem in a wireless sensor network. *Mobile Networks and Applications*, 10(4):519–528, 2005.

[64] Hong Huo, Jianjun Qing, Tao Fang, and Nan Li. Land cover classification using local softened affine hull. *IEEE Transactions on Geoscience and Remote Sensing*, 50(11):4369–4383, 2012.

[65] Teerawat Issariyakul and Ekram Hossain. *Introduction to network simulator NS2*. Springer Science & Business Media, 2011.

[66] Raymond Austin Jarvis and Edward A Patrick. Clustering using a similarity measure based on shared near neighbors. *IEEE Transactions on Computers*, 100(11):1025–1034, 1973.

[67] Theodore Johnson and Richard Newman-Wolfe. A comparison of fast and low overhead distributed priority locks. *Journal of Parallel and Distributed Computing*, 32(1):74–89, 1996.

[68] Maria Joseph and Sreeja Ashok. Minimum spanning tree based community detection for biological data analysis. *Journal of Engineering and Applied Sciences*, 12(21):5452–5456, 2017.

[69] Piotr Juszczak, David MJ Tax, Elżbieta Pe, and Robert PW Duin. Minimum spanning tree based one-class classifier. *Neurocomputing*, 72(7):1859–1869, 2009.

[70] Michael Kallay. The complexity of incremental convex hull algorithms in $\mathbb{R}^d$. *Information Processing Letters*, 19(4):197, 1984.

[71] Tai Woo Kim, Eui Hong Kim, Joong Kwon Kim, and Tai Yun Kim. A leader election algorithm in a distributed computing system. In *Proceedings of the 5th IEEE Workshop on Future Trends of Distributed Computing Systems*, pages 481–485. IEEE, 1995.

[72] David G Kirkpatrick and Raimund Seidel. The ultimate planar convex hull algorithm? *SIAM Journal on Computing*, 15(1):287–299, 1986.

[73] Jack Koplowitz and D Jouppi. A more efficient convex hull algorithm. *Information Processing Letters*, 7(1):56–57, 1978.

[74] Abhijit Kundu, Ritika Misra, Atreyee Kar, Sagarika Debchoudhury, Shilpa Pareek, Somen Nayak, and Ratul Dey. On demand secure routing protocol using convex-hull & k-mean approach in manet. In *IEEE Annual Ubiquitous Computing, Electronics & Mobile Communication Conference (UEMCON)*, pages 1–5. IEEE, 2016.

[75] Casimir Kuratowski. Sur le problème des courbes gauches en topologie. *Fundamenta Mathematicae*, 15(1):271–283, 1930.

[76] Farid Lalem, Ahcène Bounceur, Madani Bezoui, Massinissa Saoudi, Reinhardt Euler, and Marc Sevaux. LPCN: Least polar-angle connected node algorithm to find a polygon hull in a connected Euclidean graph. *Journal of Network and Computer Applications*, 93:38–50, 2017.

[77] Farid Lalem, Rahim Kacimi, Ahcene Bounceur, and Reinhardt Euler. Boundary node failure detection in wireless sensor networks. In *IEEE International Symposium on Networks, Computers and Communications (ISNCC)*, pages 1–6. IEEE, 2016.

[78] Der-Tsai Lee and Bruce J Schachter. Two algorithms for constructing a Delaunay triangulation. *International Journal of Computer & Information Sciences*, 9(3):219–242, 1980.

[79] Philippe G LeFloch and Jean-Marc Mercier. Revisiting the method of characteristics via a convex hull algorithm. *Journal of Computational Physics*, 298:95–112, 2015.

[80] Philip Levis, Nelson Lee, Matt Welsh, and David Culler. Tossim: Accurate and scalable simulation of entire tinyos applications. In *Proceedings of the 1st International Conference on Embedded Networked Sensor Systems*, pages 126–137. ACM, 2003.

[81] Ke Li, Sam Kwong, Jingjing Cao, Miqing Li, Jinhua Zheng, and Ruimin Shen. Achieving balance between proximity and diversity in multi-objective evolutionary algorithms. *Information Sciences*, 182(1):220–242, 2012.

[82] Xiang-Yang Li, Gruia Calinescu, Peng-Jun Wan, and Yu Wang. Localized Delaunay triangulation with application in ad hoc wireless networks. *IEEE Transactions on Parallel and Distributed Systems*, 14(10):1035–1047, 2003.

[83] Nancy A Lynch. *Distributed algorithms*. Elsevier, 1996.

[84] Liam P Maguire, T Martin McGinnity, Brendan Glackin, Arfan Ghani, Ammar Belatreche, and Jim Harkin. Challenges for large-scale implementations of spiking neural networks on FPGAs. *Neurocomputing*, 71(1-3):13–29, 2007.

[85] Behnish Mann and Alex Arvavid. Message complexity of distributed algorithms revisited. In *International Conference on Parallel, Distributed and Grid Computing (PDGC)*, pages 417–422. IEEE, 2014.

[86] William B March, Parikshit Ram, and Alexander G Gray. Fast Euclidean minimum spanning tree: algorithm, analysis, and applications. In *Proceedings of the 16th ACM International Conference on Special Interest Group on Knowledge Discovery and Data Mining SIGKDD*, pages 603–612. ACM, 2010.

[87] Kamal Mehdi, Massinissa Lounis, and Ahcène Bounceur. Cupcarbon: A multi-agent and discrete event wireless sensor network design and simulation tool. In *7th International ICST Conference on Simulation Tools and Techniques, Lisbon, Portugal, 17-19 March 2014*, pages 126–131. Institute for Computer Science, Social Informatics and Telecommunications Engineering (ICST), 2014.

[88] Kurt Mehlhorn. *Data structures and algorithms 1: Sorting and Searching*, volume 1. Springer Science & Business Media, 2013.

[89] L Miller, C Heymans, TD Kitching, L Van Waerbeke, T Erben, H Hildebrandt, H Hoekstra, Y Mellier, BTP Rowe, and J Coupon. Bayesian galaxy shape measurement for weak lensing surveys–III. Application to the Canada–France–Hawaii Telescope Lensing Survey. *Monthly Notices of the Royal Astronomical Society*, 429(4):2858–2880, 2013.

[90] Adriano Moreira and Maribel Yasmina Santos. Concave hull: A k-nearest neighbours approach for the computation of the region occupied by a set of points. *INSTICC Press (Institute for Systems and Technologies of Information, Control and Communication)*, 2007.

[91] Jaroslav Nešetřil, Eva Milková, and Helena Nešetřilová. Otakar Borůvka on minimum spanning tree problem translation of both the 1926 papers, comments, history. *Discrete Mathematics*, 233(1-3):3–36, April 2001.

[92] Vilfredo Pareto and Alfred Bonnet. *Manuel d'économie politique*. Marcel Giard Paris, 1927.

[93] Jin-Seo Park and Se-Jong Oh. A new concave hull algorithm and concaveness measure for n-dimensional datasets. *Journal of Information Science and Engineering*, 28(3):587–600, 2012.

[94] David Peleg. Distributed computing. *SIAM Monographs on discrete mathematics and applications*, 5, 2000.

[95] Viet Pham, Carl Laird, and Mahmoud El-Halwagi. Convex hull discretization approach to the global optimization of pooling problems. *Industrial & Engineering Chemistry Research*, 48(4):1973–1979, 2009.

[96] S. Pierre, D. Delahaye, and S. Cafieri. Aircraft trajectory planning by artificial evolution and convex hull generations. In *Air Traffic Management and Systems II*, pages 49–67. Springer, 2017.

[97] Franco P. Preparata and Se June Hong. Convex hulls of finite sets of points in two and three dimensions. *Communications of the ACM*, 20(2):87–93, 1977.

[98] Johann Radon. Mengen konvexer körper, die einen gemeinsamen punkt enthalten. *Mathematische Annalen*, 83(1):113–115, 1921.

[99] Nicola Santoro. *Design and Analysis of Distributed Algorithms*, volume 56. John Wiley & Sons, 2006.

[100] Massinissa Saoudi, Farid Lalem, Ahcène Bounceur, Reinhardt Euler, Abdelkader Laouid, Madani Bezoui, and Marc Sevaux. D-LPCN: A distributed least polar-angle connected node algorithm for finding the boundary of a wireless sensor network. *Ad Hoc Networks*, 56:56–71, 2017.

[101] Yoav Sasson, David Cavin, and André Schiper. Probabilistic broadcast for flooding in wireless mobile ad hoc networks. In *IEEE Wireless Communications and Networking, WCNC 2003*, volume 2, pages 1124–1130, 2003.

[102] Alexander Schrijver. *Combinatorial optimization: polyhedra and efficiency*, volume 24. Springer Science & Business Media, 2002.

[103] Robert Sedgewick and Jeffrey Scott Vitter. Shortest paths in Euclidean graphs. *Algorithmica*, 1(1-4):31–48, 1986.

[104] Gyula Simon, Peter Volgyesi, Miklós Maróti, and Akos Ledeczi. Simulation-based optimization of communication protocols for large-scale wireless sensor networks. In *IEEE Aerospace Conference*, volume 3, pages 31339–31346, 2003.

[105] Jack Sklansky. Measuring concavity on a rectangular mosaic. *IEEE Transactions on Computers*, 100(12):1355–1364, 1972.

[106] Sherman K Stein. Convex maps. *Proceedings of the American Mathematical Society*, 2(3):464–466, 1951.

[107] Ivan Stojmenovic, Anand Prakash Ruhil, and DK Lobiyal. Voronoï diagram and convex hull based geocasting and routing in wireless networks. *Wireless Communications and Mobile Computing*, 6(2):247–258, 2006.

[108] G Subramanian, VVS Raveendra, and M Gopalakrishna Kamath. Robust boundary triangulation and Delaunay triangulation of arbitrary planar domains. *International Journal for Numerical Methods in Engineering*, 37(10):1779–1789, 1994.

[109] Andrew S Tanenbaum. Computer networks. 4th. *Prentice Hall PTR.* *950*, 2002.

[110] Gerard Tel. *Introduction to distributed algorithms.* Cambridge University Press, 2000.

[111] Ben L Titzer, Daniel K Lee, and Jens Palsberg. Avrora: Scalable sensor network simulation with precise timing. In *Fourth International Symposium on Information Processing in Sensor Networks, IPSN 2005*, pages 477–482. IEEE, 2005.

[112] Godfried T Toussaint and David Avis. On a convex hull algorithm for polygons and its application to triangulation problems. *Pattern Recognition*, 15(1):23–29, 1982.

[113] András Varga. Using the OMNeT++ discrete event simulation system in education. *IEEE Transactions on Education*, 42(4):11, 1999.

[114] Remco C Veltkamp and Michiel Hagedoorn. State of the art in shape matching. In *Principles of Visual Information Retrieval*, pages 87–119. Springer, 2001.

[115] Klaus Wagner. Bemerkungen zum Vierfarbenproblem. *Jahresbericht der Deutschen Mathematiker-Vereinigung*, 46:26–32, 1936.

[116] Shaohua Wan and JK Aggarwal. Robust object recognition in rgb-d egocentric videos based on sparse affine hull kernel. In *Proceedings of the IEEE Conference on Computer Vision and Pattern Recognition Workshops*, pages 97–104, 2015.

[117] Bin Wang, Qinghua Ding, Xiouhua Fu, In-Sik Kang, Kyung Jin, J Shukla, and Francisco Doblas-Reyes. Fundamental challenge in simulation and prediction of summer monsoon rainfall. *Geophysical Research Letters*, 32(15), 2005.

[118] Jin Wang, Yiquan Cao, Jiayi Cao, Huan Ji, and Xiaofeng Yu. Energy-balanced unequal clustering routing algorithm for wireless sensor networks. In *International Conference on Computer Science and its Applications*, pages 352–359. Springer, 2016.

[119] Jun Wang, Hanzi Wang, and Wan-Lei Zhao. Affine hull based target representation for visual tracking. *Journal of Visual Communication and Image Representation*, 30:266–276, 2015.

[120] Jun Wang, Yuanyun Wang, Chengzhi Deng, and Shengqian Wang. Robust visual tracking based on convex hull with emd-l1. *Pattern Recognition and Image Analysis*, 28(1):44–52, 2018.

[121] Ying Xu, Victor Olman, and Dong Xu. Clustering gene expression data using a graph-theoretic approach: an application of minimum spanning trees. *Bioinformatics*, 18(4):536–545, 2002.

[122] Ying Xu and Edward C Uberbacher. 2D image segmentation using minimum spanning trees. *Image and Vision Computing*, 15(1):47–57, 1997.

[123] Boting Yang. Euclidean chains and their shortcuts. In *5th International Conference on Combinatorial Optimization and Applications, COCOA 2011, Zhangjiajie, China*, pages 141–155. Springer, 2011.

[124] Li Yang. Building k edge-disjoint spanning trees of minimum total length for isometric data embedding. *IEEE Transactions on Pattern Analysis and Machine Intelligence*, 27(10):1680–1683, 2005.

[125] Charles T Zahn. Graph-theoretical methods for detecting and describing gestalt clusters. *IEEE Transactions on Computers*, 100(1):68–86, 1971.

[126] Caiming Zhong, Mikko Malinen, Duoqian Miao, and Pasi Fränti. A fast minimum spanning tree algorithm based on k-means. *Information Sciences*, 295:1–17, 2015.

[127] Caiming Zhong, Duoqian Miao, and Pasi Fränti. Minimum spanning tree based split-and-merge: A hierarchical clustering method. *Information Sciences*, 181(16):3397–3410, 2011.

[128] Caiming Zhong, Duoqian Miao, and Ruizhi Wang. A graph-theoretical clustering method based on two rounds of minimum spanning trees. *Pattern Recognition*, 43(3):752–766, 2010.

[129] Weiqiang Zhou and Hong Yan. Alpha shape and Delaunay triangulation in studies of protein-related interactions. *Briefings in Bioinformatics*, 15(1):54–64, 2012.

Index